AF680742

Environmental Technology for Architects

Environmental Technology for Architects

Volume 1: Comfort, Context, and the Sun

First Edition

Martin Gold, Gary Siebein, Keely M. Siebein, and Sara Bensalem

University of Florida

cognella
SAN DIEGO

Bassim Hamadeh, CEO and Publisher
Seidy Cruz, Specialist Acquisitions Editor
Danielle Gradisher, Senior Project Editor
Susana Christie, Senior Developmental Editor
Rachel Mann, Senior Instructional Designer
Samantha Hansen, Production Editor
Jess Estrella, Senior Graphic Designer
Alexa Lucido, Senior Licensing Manager
Natalie Piccotti, Director of Marketing
Kassie Graves, Senior Vice President, Editorial
Alia Bales, Director, Project Editorial and Production

Copyright © 2027 by Cognella, Inc. All rights reserved. No part of this publication may be reprinted, reproduced, transmitted, or utilized in any form or by any electronic, mechanical, or other means, now known or hereafter invented, including photocopying, microfilming, and recording, or in any information retrieval system without the written permission of Cognella, Inc. For inquiries regarding permissions, translations, foreign rights, audio rights, and any other forms of reproduction, please contact the Cognella Licensing Department at rights@cognella.com. While every effort has been made to ensure the accuracy of this work, the publisher and author(s) assume no responsibility for errors or for the consequences of their use, and makes no warranties, express or implied.

Trademark Notice: Product or corporate names may be trademarks or registered trademarks and are used only for identification and explanation without intent to infringe.

Printed in the United States of America.

cognella® | ACADEMIC PUBLISHING
320 South Cedros Ave., Ste. 400, Solana Beach, CA 92075

Contents

Contents

Foreword

Martin Gold

The authors of this "electronic book" offer an initial and perhaps transitional mode of learning that supports both traditional lecture-based support course models and design studio learning. Offering insights and perspectives accumulated over a combined 60 years of award-winning educating and practice in architecture and consulting, the authors seek to advance architectural pedagogy. Relying on the vast resources and almost instantaneous access of the internet, emergent communication and resource platforms, and the prolific use of digital drawing and modeling, this set of learning modules seeks to provide foundational theories, concepts, and applied examples in the environmental technologies that support learning as a tool that can be integrated into the architectural design studio.

Modes of communication, design, and access to information have been revolutionized by digital technology during the transition from the 20th to the 21st century. Education, although slow to transition, was catalyzed in 2020 by the COVID-19 worldwide pandemic propelling the "digital meeting" into the forefront as the primary mode of interaction and subsequently challenging the definition of "face-to-face" as a physical necessity. In architectural design studios, digital pin-up walls or "white boards" have replaced the physical walls, ironically making the content more accessible. By eliminating the need to print and pin-up digital materials, and by utilizing common cell phone cameras to image sketches and physical models, the "digital wall" gives everyone a "front row seat" to the creative content and critique. The mode invites everyone in the classroom to participate in real-time mark-ups, post comments, paste images of precedents, and add links to support materials emerging during the critique. These tools cannot replace the activity of doing the intuitive sketch, thumbnail diagram, bug model, or the critical thinking that happens at a student's desk as others work alongside in the studio. However, this is truly a revolutionary change in the process of educating designers. And, the records of those intense and fluid moments of discourse are archived indefinitely rather than being taken away at the end of the critique.

Percolating underneath these rapidly emerging modes of design is the call for collaboration between design and implementation professions. For over two decades, the idea that electronic documents could allow the many designers, consultants, and constructors to participate directly throughout the design and delivery process has been written about, debated, and incorporated into the learning objectives of accreditation bodies and, more recently, the licensing bodies including the NAAB, ACCE, CIDA, NCARB, CCM, NCIDQ.

In terms of collaboration, the professions are leading the way offering an array of architectural design and delivery modes, including Integrated Project Delivery (IPD), Design-Build (DB), and Guaranteed Maximum Price (GMP), to name a few. These modes of architectural practice rely heavily on early and continued collaboration between diverse project teams bringing expertise and knowledge to bear in a creative and collaborative environment. For architects, in particular, this demands an evolving role in the process that might be characterized as transitioning from a dictatorial to a transformational leader. Rather than allocating tasks, these leaders seek to establish outcomes, collaborate, encourage, and inspire teams toward a collectively developed vision. It must be clarified that this mode is quite different from the so-called "over the wall" design modes of tossing the drawings to the next group of consultants to "engineer" lighting, HVAC, or life safety as after-the-fact retrofits tacked onto the architect's design vision. Nor does one expect the architect to acquire all the knowledge necessary to manage the complexity of the buildings of our age. What is required of architects is a clear and inspiring architectural vision and the skills to bring consultants together around that vision. Architects need a clear understanding of technological principles, techniques of application, and an understanding of what information consultants need to contribute toward and achieve the vision.

Recognizing these emerging circumstances, proactive Schools of Architecture are reconsidering the traditional dominance of the sole designer studio model that is typically underpinned by separate "support courses"—courses that cover expertise from related consultants in areas such as materials, structures, mechanical, lighting, acoustics, etc. Architecture schools in the United States are asking an important question: How can we embed broader knowledge, appropriate for architects, as an integral part of the studio mode of educating students of architecture and their allied disciplines?

This series of modules seeks to chart new territory in that direction by offering insights and interactive content toward educating students to the "be able to" level of implementing appropriate knowledge of environmental technologies in their designs. Students will be asked to deeply engage the global community of relevant digital information as their source for timely technical information. Learning modules will seek to frame critical thinking by introducing theoretical positions, conceptual approaches, and practical strategies that have been used successfully in noted projects. Sustainability, resiliency, and natural ecology, as a few examples, each offer important theoretical, conceptual, and applied issues that are critical to the future of humanity and to the knowledge base of emerging architects. Here, the authors seek to frame this knowledge in a manner that is concise, accessible, and that can be explored, understood, and implemented by students at the early schematic design level in academic design studios.

Module 1

Introduction to Environmental Technology

Martin Gold

We shape our buildings; thereafter they shape us.

—Winston Churchill

Churchill's famous comment was directed at the shape of the British Commons Chamber after it was destroyed by Nazi incendiary bombing during World War II in 1943 (UK Parliament, 2019). Churchill argued that the historic rectangular form rather than a proposed new semi-circular form—was fundamental to the adversarial two-party system of a successful parliamentary democracy. His expression has often been extended more broadly to highlight the critical impact of our built environments on who we are as individuals and societies and whom we might become. Gaston Bachelard, in his highly influential text, *The Poetics of Space,* introduces us to the fundamental relationships between our childhood homes and those experiences of: the cold, dark, and damp basement; the creaky and breezy attic; the amber warmth and crackle radiating from the hearth; and the smells that percolate through the air as food is cooking in the kitchen. Our individuality and our poetic framing of the world around us, are constructed from the accumulation of these experiences as we dwell in spaces and in the landscape. As students of architecture, we must seek to understand the ecology of these relationships as systems of physical conditions, three-dimensional space, perception, and interpretation. As architects, we are responsible for visioning the next generation of, and creating these formative environments. Environments that James Marston Fitch refers to broadly as the "third environment" "interposed between man and nature." If we are successful as architects and designers, we will have managed to connect people to the beauty and temporality of the natural environment in an inspiring and healthful manner while also minimizing the many stresses imposed by the natural environment. We will also use resources carefully and in moderation. These fundamental concepts had not changed since the 1st century BC when author and architect Marcus Vitruvius Pollio (Vitruvius) expressed three fundamental components of built space—*firmitas, utilitas, and venustas*—often translated as firmness, commodity, and delight. But perhaps those elements are just the essentials. Renowned Finish architect Alvar Aalto, suggested, during a 1957 lecture, that architects "wish to build a paradise on earth for people."

Figure 1.1 Housing for Factory Workers at the Sunila Pulp Mill, Kotka, Finland; Alavar Aalto Seeks to Humbly Connect Workers With Light, View, and Nature as Both Amenities and Environmental Aspects of Healthful and Rewarding Living.

Figure 1.2 Hyvän Paimenen Kirkko, Helsinki, Juha Leiviska; the Altar Becomes a Choreography of Light, Sound, and Temporal Change, Protecting From the Harsh Nordic Winter While Embracing the Light of the Sky and the Verticality of the Adjacent Birch Forest.

Architecture is a challenging profession in any era but especially in the early 21st century given the array of: specialized experts; a plethora of emergent materials; dynamically advancing fabrication modes; sophisticated form bending computer modeling and measurement techniques; advancing expectations for comfort and efficiency; direct access to mind-numbing amounts of information; the emergence of Artificial Intelligence (AI) to somehow manage vast amounts of information; and a variety of project architecture delivery modes that have emerged in the early 2100s. Clients are more often steering committees, boards of directors, or communities rather than individuals. And, even as individuals, they may not have a clear idea of what they want from their architecture. The role of the architect is also radically evolving. Students will enter a profession that requires considerable generalized knowledge that will be deployed in detail by specialized consultants, engineers, and construction managers. Each consultant will endeavor to optimize their domain in the project (and rightly so). The architect must direct these many perspectives in a manner that optimizes the design for a specific holistic result. Almost all projects are largely unique, and "optimal" design may require the appropriate subordination of many competing independent optimizations. In other words, no single aspect of a complex project will likely be optimized, yet many complimentary elements will be well-considered in a successful project. Architecture requires the resolution of hundreds to thousands of contradictory forces that must be brought into harmony. "This harmony cannot be achieved by any other means than those of art."

Figure 1.3 A Pair of Cardinals Have Sited and Constructed a Nest for Their Family of Five in the Shade and Rain Shelter Under Large Fig Leaves, Strategically Nestled in Between Structural Branches, and a Nursery Space Formed From the Available Materials of Sticks, Dried Grasses, and Spanish Moss.

The collection of knowledge modules put forward in this series offers theoretical perspectives on the relationships between humans, their senses, and their environments. Conceptual modes of thinking are presented that prioritize the interrelationships and connectivity among nature, technology, humanity, and architecture. Practical strategies developed and deployed in the distant and recent past are presented to give students foundational knowledge that should evolve as their professional judgment and confidence develop through their experience practicing the art of architecture.

Technology

Defining the boundaries of environmental technology can be elusive. It could be considered *the interface among the dynamic natural environment, variable human sensation, and the variability of the psychology of perception*. We might offer that Environmental Technology is the range of techniques people have developed over the years to control their environments and to make their dwellings more inspiring, useful, and comfortable. The philosopher Martin Heidegger offers an etymology of "technology" as being comprised of legin (-logy), to speak of, and apophainesthia (techne), to bring forward into appearance. Heidegger's definition of "bringing forward" is quite interesting, in that it draws from his perspective that all things to be "discovered" already exist—they just have not yet been revealed. For Heidegger, technology is a process of revealing or discovery (rather than something fabricated). He also suggests this revealing is not a technological process but rather a creative process. He argues that the essence of technology is largely ambiguous and a paradox in that "exact science" could "enframe" us and prevent the process of revealing (creativity)—only through modes of art and poetry can this paradox be resolved.

The essence of technology is by no means technological.

—Martin Heidegger

In considering Heidegger's proposition, we can also consider three levels of technology: theoretical, conceptual, and practical. As a theory, we might ask, what is the role of technology in architecture? Le Corbusier (Corbu) draws the sun as a hybrid of mythical Greek figures Aphrodite and Medusa. We can interpret this as theoretical paradox. For humans, the sun might be warm and beautiful or oppressive, inhumane, and petrifying. Of course, it is both, and the next consideration is how this phenomenon might translate to architecture. Corbu responds with a concept—the brise-soleil—literally translating to "breaks the sun." It is clear from Corbu's diagrams that the brise-soleil is more complex than the direct translation implies. He is seeking to embrace the warmth and beauty of the sun during the winter while deflecting the harsh heat and light during the summer.

Figure 1.4 LeCorbusier's Juxtaposition of Aphrodite and Medusa Teaches Us That the Sun Can Be a Beautiful Warming Presence Brightening an Otherwise Dreary Day Yet Also Having the Intensity of Heat and Glare on a Hot Summer Day to Turn Us to Stone. The Paradoxical Power of the Sun to Warm Us or Bake Us, Must Be Well Understood by Architects.

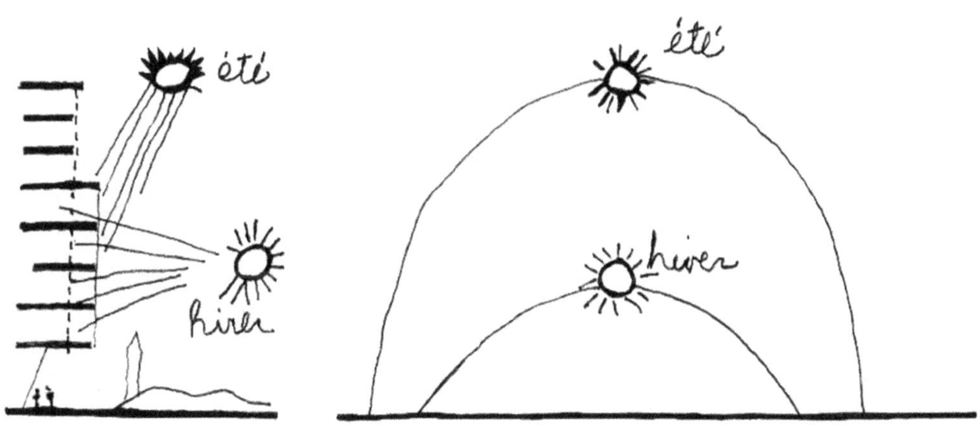

Figure 1.5 LeCorbusier's Conceptualization of the brise-soleil as a Physical Response to the Seasonal Position of the Sun; the Bold Horizontal Lines on the Left Show How the Shorter and Longer Balconies Allow the Sun's Rays to Penetrate More During the Winter (*hiver*) Months and Provide Shading During the Summer (*été*); on the Right, LeCorbusier Also Provided the Sun's Arcs During the Winter and Summer Months so Viewers Could Extrapolate the Cause of the Difference in Sun Angle From Winter to Summer and Apply It in Their Own Work.

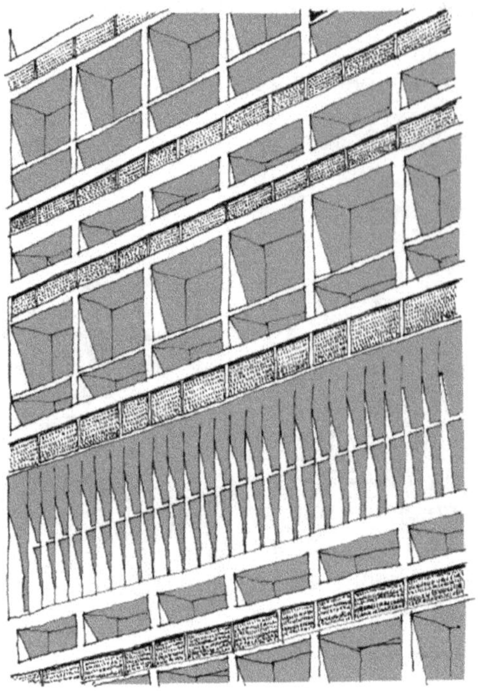

Figure 1.6 Shade Study of LeCorbusier's brise-soleil Concept Applied to the Unite d' Habitation Housing Project, Marseille, France, 1952.

Corbu's brise-soleil represents a conceptual idea that is drawn from a theoretical understanding of the solar path and radiant heat. These principles are applied, in this example case, as an implementation of the brise-soleil as architecture. Corbu, and many other architects, have employed the brise-soleil as an architectural system that also responds to and perhaps enhances views, natural air flow, protective railings, and heat sinks as part of providing occupied spaces. The theory is translated through concepts appropriate to architecture as part of a design process considering the relationship among the sun, the building façade (orientation), climate, building program, materials, and of course budget.

The Environment

When we speak of the environment, we must consider the totality of that which surrounds us (viron), French – circle, to turn, or around) (Environ – Wiktionary, the Free Dictionary, 2022). The Oxford English Dictionary defines environment as the totality of that which determines, modifies, and influences the character and quality of life. The philosopher Jose Ortega y Gasset describes the natural environment as "…an intricate net woven of both facilities and difficulties." In the case of the natural environment, we are considering the world that has not been altered by human civilization, the world of the out-of-doors, so to speak. The world that humanity first encountered and responded to with ingenuity, creativity, and technology. These "facilities and difficulties" come to us as a range of influences such as: warmth-heat, cool-cold, arid-humid, windy-balmy, rain-sunshine, etc. LeCorbusier's understanding of the sun as both facility and difficulty, led him to understand there is an opportunity for design, and a role for environmentally responsive technology in architecture.

Within the ranges of stresses occurring in the vastness of the natural environment, we experience and begin to define the limits of habitation. Those stresses that are tolerable, and the even narrower range of conditions that offer human comfort and perhaps even joy, must be nurtured with our architecture. We have all experienced the satisfaction of the warming sun on a chilly day or the relief of a cool breeze or the shade on a sunny day that may be getting just too warm. It is within these relatively narrow ranges of environmental conditions that we achieve comfort—a condition that allows modern humanity to flourish.

Figure 1.7 Garden at the Calligraphy Museum, Xi'an, China; These Asian Gardens Have Inspired Many Architects, Including Frank Lloyd Wright, to Celebrate the Connection Between Architecture and the Natural Environment—Linking Our Sheltered Spaces With "That Which Surrounds".

Figure 1.8 Winter in Vals, Switzerland; a Picturesque Landscape, Yet It Requires an Infrastructure of Dwelling to Survive the Cold and Barren Landscape.

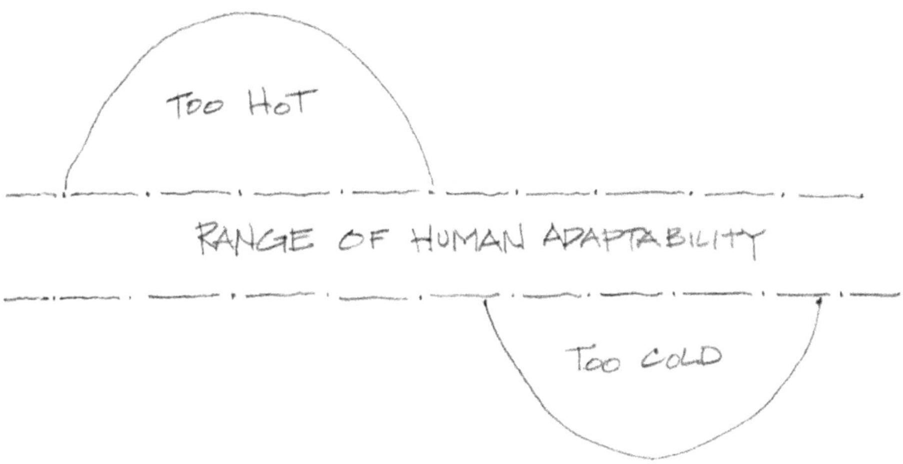

Figure 1.9 Range of Human Adaptability Diagram. Environmental Stresses, Such as Heat or Cold, Outside Our Range of Adaptability, Begin to Disrupt Both Our Intellectual Capacity and Our Bodily Functions and, in Prolonged Cases, Can be Fatal.

Jose Ortega y Gasset proposes that "technology is man's reaction upon nature or circumstance. It leads to the construction of a new nature, a super nature, interposed between man and original nature." He goes on to discuss the evolution of this "super nature," created by man, that is driven by desires rather than needs. In other words, our quest to control our environment is driven by aesthetic intention, rather than by a reactive response. In this aesthetic process, people are motivated by desires rather than needs: rather than walk, we train the horse, we invent the automobile, and eventually, we discover the technology of flight and craft the airplane. We have constructed rockets that have put people on the moon yet there was clearly no inherent need to do that. Ortega y Gasset describes three phases of technology: 1. Accidental invention, 2. Craftsmanship and tool making, 3. Technician—individually operative experts.

Here we should clarify the two important concepts of objectivity and subjectivity. Although technology, as proposed above is driven by aesthetic desires rather than needs, the measurement and the scientific study of objective reality is critical to advancing technology. On the other hand, subjective interpretation and prioritization are also critical to assess the benefits of technology. Gaston Bachelard writes that objective reality is inherently scientific—"what it is as measured"—and subjective reality is inherently human, asking "what it means" to people. *Environmental technology, as a fundamental element of architecture, must reconcile these seemingly divergent measures of reality—objective and subjective—as a fundamental aspect of the architectural design process.*

Architecture, The Third Environment

In 1972, Author James Marston Fitch introduced the concept of architecture as a "third environment interposed between man and nature." Fitch describes our human metabolism—our bodies—as the first "micro" environment. Humans and animals interact with the "macro" or natural environment through the interface of the skin, hair, and clothes. That interface is regulated by our metabolism and the body's ability to accommodate changes in activity and the conditions or stresses of the macro environment.

Fitch argues that architecture creates a new "meso" (middle) environment interposed between humans (micro) and the natural (macro) environment. In this meso environment of architecture, the range of natural stresses on humans can be moderated, mitigated, nurtured, or enhanced. In this theoretical model—the techniques humans have developed to make architecture—become an interface between this new meso environment and the natural or macro terrestrial environment. Technology, by extension, then becomes the primary interface between humans and the natural environment.

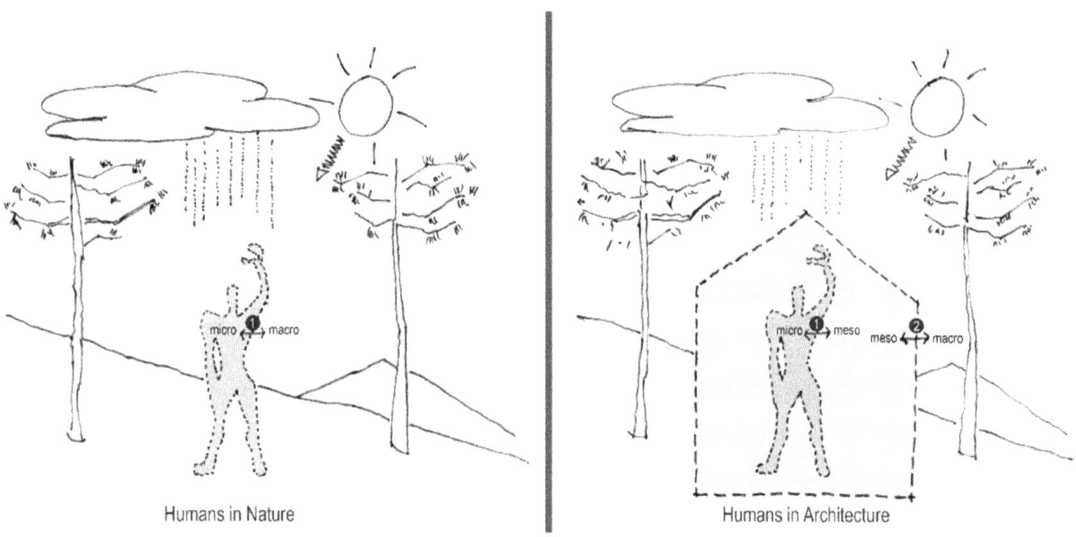

Figure 1.10 Architecture as the Third Environment, Interposed Between Humans and the Natural Environment; Interface 1 Is the Exchange Between Human Skin/Clothes and the Natural Environment—Biological; Interface 2 Is the Exchange Between Architecture and the Natural Environment—Technological.

Technology as Interface

The diagram below illustrates some of the complex relationships between Fitch's meso environment and the natural environment. In this relatively new modern environment of architecture, human beings have developed sophisticated social and cultural sets of activities—dwelling, industrial production, office production, worship, theater, education, laboratory, studio, hospital, warehouse, retail commerce, dining, and others. Each of these program activities has specific environmental requirements that are typically within a narrow range of those provided by the natural environment, which of course, changes with global location. Certainly, on a beautiful day, we can perform most of these activities out-of-doors and enjoy them when we can. However, most of the year, in most places, regulated environments are necessary.

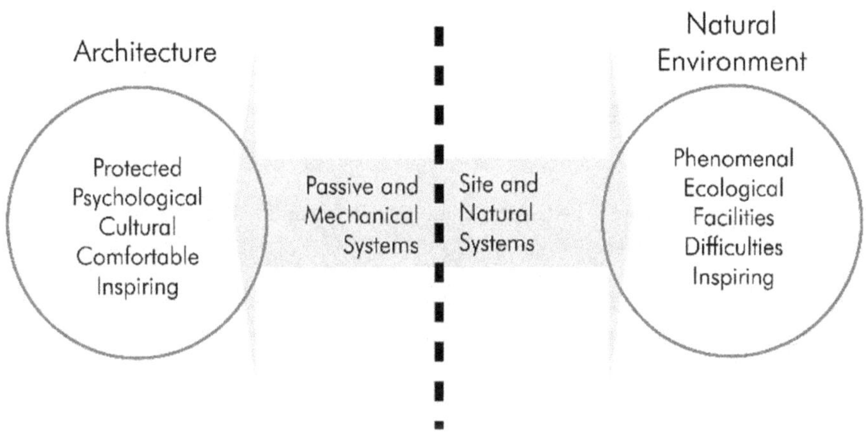

Figure 1.11 Technology as the Interface Between Architecture and the Natural Environment.

As an interface between the meso (architecture) and macro (natural environment), the technology that architects employ, must draw on the facilities of the natural environment and reduce the difficulties of creating spaces that are both inspiring, comfortable, and useful. People must be connected to the phenomenal character of their environments and linked to the ecologies of their localities while being supported and protected both physically and psychologically. People must feel comfortable within an appropriate range of environmental stimulation and should have a positive sense of well-being.

A Tale of Two Architectures

Perhaps we could consider two radically different approaches to creating a meso environment. In the first case, we would use no energy, and the architectural interface would be constructed from things readily available around us. In fact, almost all architecture was constructed this way prior to 1930. In the example below, the Seminole Indian "chikee" (translates to house) is fabulously adapted to a hot, humid, subtropical climate. The palm thatch roof provides shade protection from the intense sunlight, while the lack of walls maximizes airflow capturing the light breezes. The palm thatch is very lightweight, and so it sheds its heat very quickly after the sun sets. The floor is raised off the ground to maintain a dry, well-ventilated surface, as the ground typically remains moist or wet after rains. The orientation might have the gable ends facing east and west allowing morning sun to enter in the east, warming the space from a cool night, while strategically locating the structure adjacent to a clump of palm trees on the west side to provide shade from the low afternoon sun. We refer to these strategies as "passive" as they do not require any energy input to mitigate the stresses of the natural environment.

Figure 1.12 A Tale of Two Architectures. On the Left, Is the Ubiquitous Refrigerator, Closed and Sealed; on the Right, a Sketch of a Seminole Indian Chikee, Open and Connected.

In contrast, we could consider the optimal climate-controlled environment—the refrigerator. It is well insulated, limits air exchanges with the outside to the extent possible with tight seals, and has no windows as they easily transmit heat. The interior is only illuminated when opened, and the space is set to a fixed and constant temperature. Of course, we do not live in refrigerators, yet the principles that drive the design of a refrigerator are the same that drive the design of air-conditioned buildings—limit air exchanges (leaks), limit windows to avoid heat gain (or loss), set one fixed temperature, ensure the temperature is low enough to control humidity, and maximize the area of opaque highly insulating materials. We refer to this primary strategy "refrigeration" as an active strategy because it solely relies on mechanical systems and energy input to mitigate the stresses of the natural environment.

Figure 1.13 The Architectural Expression of Two Architectures. On the Left, the Portland Municipal Offices (Michael Graves, 1993), Closed and Sealed; on the Right, the Bateson Building Housing Municipal Offices in Sacramento, California (Sym van der Ryn, 1977), Open and Connected.

Our architecture must find an appropriate balance between the two extremes of controlling the thermal environment in buildings: (1) dominated by active mechanical systems requiring large energy inputs and producing thermally static spaces; or (2) controlling the thermal environment passively, relating to the context, and minimizing the external energy required. We could extend this lesson to light, sound, moisture, fresh air, water, and waste systems and will do so in the modules that follow.

Priority of Environmental Strategies

When we think of environmental stress in architecture, we are concerned when they are outside the range of human comfort for a particular activity, a particular age group, or a given climate. If the exterior environmental conditions are far outside that range, then the architecture and the building systems must respond to bring the stress within the range of comfort. Conceptually, there are three strategies that can be deployed in any given site or climate to moderate the thermal environment.

Natural Systems such as how a building is sited on the land, or in the land, or if trees or vegetation can be planted to shade the building or act as a buffer to cold winter winds, or by planting deciduous trees to the south to allow natural warming from the sun in winter and to shade in summer. Natural systems strategies can be employed to ultimately reduce the stresses on the interior spaces of the building.

Passive Architectural Systems such as overhangs, shade screens, or double skins that provide shade; or high thermal mass materials to mitigate large temperature swings; or highly insulating materials that can avoid excess heat loss or gain without using energy.

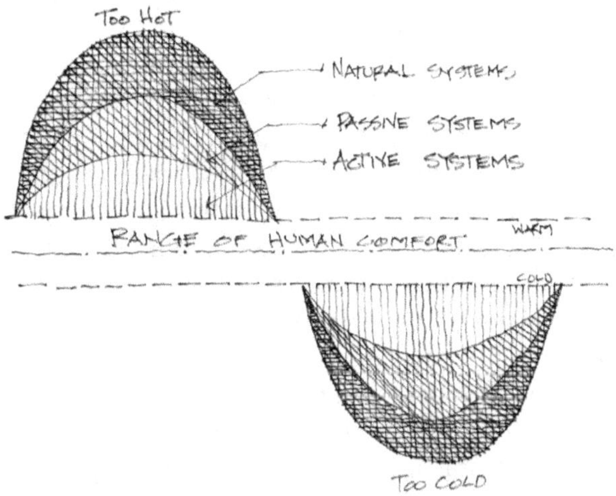

Figure 1.14 Prioritization Diagram for Mitigating Environmental Stresses—Natural, Passive, Then Active Systems

Active Mechanical Systems such as heating, cooling, dehumidification, and mechanical ventilation systems should be utilized to mitigate any remaining environmental stress. There is also a great opportunity within mechanical system selection in terms of ground-coupled heat exchange or low-energy evaporative cooling to provide effective comfort control that is responsive to the potentials of the site and climate to limit external energy use.

For architecture to become more sustainable and use less energy to maintain comfort, the priority of design considerations should first consider natural systems, then passive architectural systems, and then mechanical systems. In this fashion, the heat or cooling load on the building is minimized to the extent possible and then the mechanical systems are designed for this reduced heating or cooling load. Furthermore, optimizing natural and passive systems will extend the times of year that are comfortable without the need for any cooling or heating.

The Evolution of Architectural Technology

At the beginning of the 19th century, approximately 20% (1 in 5 people) of the U.S. population worked directly in agriculture to feed the total population of approximately 17 million people. This lifestyle was mostly lived out-of-doors and during daylight hours. By 2015, only 0.6% (1 in 166 people) of the population worked in agriculture to feed approximately 331 million Americans. In 1925, only about half of U.S. homes were supplied with electricity. Manhattan, for example, was dominated by farmland in the early 1800s, with very little manufacturing and almost no commercial office space. Goods trading was the primary activity. By the beginning of the 21st century, Manhattan became a megalopolis of predominantly office spaces, housing, and retail activities. As of 2023, very little shipping or manufacturing remains in Manhattan which once dominated the waterfront. As humanity has migrated from the fields, through manufacturing, and on to the more recent dominance of office spaces and service work, technology also has advanced and adapted to support these changes.

This evolution of building programs and the associated technology has become highly complex in the 21st century, requiring computer-based modeling and design software (e.g., CAD, BIM); collaboratives of specialized technical experts; responses to challenging code requirements for safety and the flow of business; and the evolution in the methods of design, construction, and building delivery.

The COVID-19 global pandemic that produced a temporary migration of workers to "work from home" seems to be productive, avoided costly commutes, and took advantage of untapped digital technologies—the virtual meeting space. As the pandemic ebbed, corporations established a new normal with hybrid working models, less time/space in offices, and more reliance on the home as a viable workspace. Additionally, "brick and mortar" physical retail environments have been significantly challenged by online retailers and then disrupted again by COVID-19. As early as 2009, the term "dead mall" was coined to describe this phenomenon occurring across suburban America. These transformations are occurring in large, medium, and small cities around the world and will drive the next generation of urban dwellings, suburban homes, and the transformation of outdated office buildings. The technology to heat, cool, light, ventilate, and quiet these new and repurposed spaces will also evolve and change.

One emerging developmental model—mixed-use—offers to draw from historic pre-automobile cities to "re-stitch" the activities of live-work-play back together in smaller modular communities. This strategy can scale from low-rise communities to skyscrapers in urban areas. However, this complexity will demand emergent technologies that seek to bring the natural world into high-density living, reintroduce agriculture and food production into the city, and will require strategies to maintain comfort and privacy within noisy and vibrant contexts. Densely populated cities will require a dramatic philosophical shift in the understanding of resource loops and how to close them—using and reusing what is already available at that place—within both communities and buildings. Our thinking, and subsequent technology, will need to address the transient nature of building in coastal areas by asking the question: How will we utilize the existing coastal urban infrastructure as long as possible while knowing we must leave it in the near future?

For architects to operate within these dynamic conditions, technology must be understood philosophically, with a foundation in traditional concepts of objective physical reality, and with a strong sensibility for the subjective reality of aesthetics and interpretation in which humanity dwells. These environmental technology educational modules offer insights and perhaps a foundation toward that goal. Students of architecture should carry these ideas forward while continually challenging the modes and techniques of application to knowledgeably work with; educate; and learn from consultants and clients toward inspiring architecture.

> *Architects who have aimed at acquiring manual skill without scholarship have never been able to reach a position of authority to correspond to their pains, while those who relied only upon theories and scholarship were obviously hunting the shadow, not the substance. But those who have a thorough knowledge of both, like men armed at all points, have the sooner attained their object and carried authority with them.*
>
> *—Vitruvius (c. 70 B.C.–c. 25 B.C.)*

References

Aalto, A. (1972). Arkkitehtien paratiisiajatus [The paradise idea of architects]. In G. Schildt (Ed.), *Alvar Aalto: Luonnoksia* (pp. 101–102). Otava. (Lecture given in Malmö, Sweden, 1957).

Aalto, A. (1972). Taide ja tekniikka [Art and Technology] (J. Pallasmaa, Trans.) In G. Schildt (Ed.), *Luonnoksia: Alvar Aalto* (pp. 87–88). Academy of Finland. (Lecture, October 3, 1955).

environ – Wiktionary, the free dictionary. (2022). Wiktionary. https://en.wiktionary.org/wiki/environ

Heidegger, M. (1954). *The question concerning technology*. Garland Publishing.

Ortega y Gasset, J. (1939). Man the technician (H. Weyl, Trans.) In *Toward a philosophy of history*. W.W. Norton & Company.

Our World in Data. (2021) *Various sources, but mainly Groningen Growth and Development Centre 10-Sector Database*. Our World in Data.

Pollio, V. (June 1, 1960). *Ten books on architecture* (M. H. Morgan, Trans.). Dover Publications.

UK Parliament. (2019). *Churchill and the Commons Chamber*. UK Parliament. https://www.parliament.uk/about/living-heritage/building/palace/architecture/palacestructure/churchill/

Vitruvius DeArchitectura translation.

Image Credits

Fig. 1.4: Copyright © 1945 by Le Corbusier.

Fig. 1.5: Le Corbusier, "Brise-Soleil Concept," https://www.architectural-review.com/archive/le-corbusier-in-the-sun. Copyright © 1936 by Le Corbusier.

Module 2

People and Their Senses

Perception and Cognition

Martin Gold

Thinking is more interesting than knowing, but less interesting than seeing.

—*J.W. Goethe*

Introduction

How do we "see" the world around us? In his highly influential book, *The Eyes of the Skin: Architecture and the Senses*, Juhani Pallasmaa argues that architecture has become dominated by the visual realm prioritized by late 20th Century technological and consumer culture. And, the lack of consideration of our full range of senses—as an integral part of the architectural experience—has impoverished our environment and led to a culture of detachment and alienation. Pallasmaa references The Incredulity of St. Thomas by Caravaggio (the original "doubting Thomas") who would not believe Jesus was resurrected until he could touch his wounds—seeing was not believing.

Figure 2.1 The Incredulity of St. Thomas, Caravaggio, 1601–1602; Seeing Is Not Necessarily Believing and We Rely on All Our Senses When We Experience Architecture.

These themes are at the core of environmental technology theory and practice. Our worlds—natural, urban, and architectural—are multi-sensory, temporal, dynamic, and participatory environments that must be considered beyond the visual organization of plan, proportion, historic memes, or stylistic interpretations. Those architectural tools are surely important, but those architects who have a fuller sense of architecture, as layered sensory experience, and have the understanding of these interactions as both sensory and interpretive, will rise to the top of the profession.

The Experience of Architecture

Figure 2.2 Grand Bazaar, Istanbul, Turkey. A Complete Experience of Sights, Sounds, Smells, and Spaces Expanding and Compressing as One Moves Through the Maze of Shops and Courts.

Istanbul's Grand Bazaar is a great example of an architecture of experience. The sounds of musical instruments such as the baglama, davul, kaval, ud, or zurna ring out and mix in the air with the aromas of spices like oregano, cinnamon, cumin, sumac, and freshly made baklava. Tourists and shopkeepers speak multiple European, Asian, and African languages and dialects. One feels the subtle shifts in slope and the compression/expansion of space moving through the varied archways, barrel vaults, and domed spaces that have evolved over ages. Hookahs offer other aromas and leave a slight haze in the labyrinthian but open-air corridors. The intensity of these sensations is heightened rather than mitigated by the architecture. Hard tiled surfaces and domed or vaulted ceilings concentrate the sound reflecting it back to the occupants as it is both diffused and refracted through the labyrinth. These same massive walls are cool in the warm air that slowly percolates into and through the spaces collecting and redistributing the many aromas. Daylight beams in through the haze with fresh air at specific clerestories or light wells and both being diffused into the darker, cooler, and more remote spaces. Even the grime in places like these becomes part of the patina of utility that makes the experience humane and authentic.

Much has been said and written about architecture that is both humane and inspiring. The work of Alvar Aalto, for example, has inspired volumes of texts defining his work as humanist architecture. At the core of his design principles is an architecture of both utility and sensation that seeks to draw out inspiration from nature, daylight, sound, fresh air, the aroma of the Finnish forest, and the tactility of those things like door handles that we literally reach out and grasp.

Architecture and the Related Arts

In every case [of creative work] one must achieve the simultaneous solution of opposites. Nearly every design task involves tens, often hundreds, sometimes thousands of different contradictory elements, which are forced into a functional harmony only by man´s will. This harmony cannot be achieved by any other means than those of art.

—Alvar Aalto, translated by Juhani Pallasmaa

Alvar Aalto has illuminated, in these two sentences, both the dilemma of, and the strategy for, architecture. We must know the measure of conflicts, have a theoretical framework, and the elasticity of poetic thinking to guide priority in complex circumstances to find harmony. No one element or priority can completely dominate. Each element must be considered in terms of its potential for connectivity to other things—what might be considered resonances or reverberance that allows voices to blend in coherent harmony. We must be able to master the power of poetry, ambiguity, and imagination. We must also understand, with precision, the physicality of our environments, and how, as humans we might sense and interpret them.

Poetry does not fly above and surmount the earth in order to escape it and hover over it. Poetry is what first brings man onto the earth, making him belong to it and thus bring him into dwelling.

—Martin Heidegger

Painting

The architect, Louis Kahn, once said, "A painter can paint square wheels on a cannon to express the futility of war. A sculptor can carve the same square wheels. But an architect must use round wheels" (attributed to Juhani Pallasmaa). We can admire the irony, pastoral content in painting or the masterful use of color, stroke, chiaroscuro, and perspective depth. Perhaps we are interested in the hyper-reality of the old masters such as Caravaggio or Jan van Eyck, or the figurative and juxtaposed paintings of the modernists such as van Gogh, Picasso, Monet, or Léger.

Figure 2.3 Monet Haystacks, End of Summer; Fixing a View, That Is a Moment in Time, To Be Revisited as a Spectator at a Later Time.

Monet's famous haystacks studies capture the seasonal light at the end of summer, simplifying the forms in light and shadow, blurring the colors of the stacks, fields, trees, and the hills beyond while using varied tonality to achieve perspective depth. But if we come to visit the painting again in a week or two, or a year or two, it offers the same moment in time. It is a two-dimensional fixed point of view, inherently static in time, and most likely it will be, if lit consistently by the museum, the same every time we visit. It is still to be admired.

Figure 2.4 David, Gian, Lorenzo Bernini, 1623–1624; a Moment, Frozen in Material and Time, Observable From Multiple Perspectives, Yet It Remains a Static Object.

Sculpture

Three-dimensional and spatial sculpture brings together form and materiality rendered in light. In the case of Gian Lorenzo Bernini's *David*, who is about to slay Goliath, David reaches out into the space of the observer. The intensity of his task and emotional focus are palpable. We can walk around David and observe the tension in his ligaments and muscles, all carefully sculpted by Bernini. The whiteness of the marble harmonizes with light and shadow to revealing the subtle detail in the body, fabrics, and David's sling. However, this is again a static experience, the statue will not change its character or position. We may see it in a different light or with a different frame of mind, but like the painting, the sculpture is fixed in its gaze, position, and moment in time.

Theater and Cinema

These art forms are closely related to architecture by presenting the spaces of cities, streets, and rooms that we understand through our experience of occupation. Much has also been written on the exchange of ideas between cinema and architecture in terms of temporal structure, the emotive daydream, and poetic or idealized images of dwelling, domesticity, or our cultural archeology—the past, present, and future us. In the theater, the presence of the audience is critical as the "oohs," "gasps," laughter, and applause, energize the actors (theater) producing a tangible experience of participation. However, we are still not active participants in these events. We are spectators. The events do not unfold based on our presence. The show will go on, it will unfold in the same temporal order, and the outcome will be the same night after night.

Figure 2.5 Theater and Cinema Unfold and Are Dynamic Over Time, but the Narrative Is Unchanged by the Observer; the Audience Are Spectators Rather Than Participants.

Music

"I say architecture is frozen music," commented Wolfgang von Goethe, perhaps ruminating on the Baroque styles of architecture and music in the 18th century that blended repeating motifs and contrapuntal rhythms. Eila Hiltunen's sculpture *Passio Musicae* (Frozen Music) seems to capture this idea visually, honoring the renowned composer Jean Sibelius in Helsinki, Finland.

Proportion and natural harmonics first emerged through music, and these discoveries have a long history in architecture, dating back to the first Greek amphitheaters and temples. When we speak of harmony in architecture, it is the blending of disparate elements into a cohesive whole, an analogy drawing on our sensation of listing to disparate voices, of varying timbre, singing different notes that blend together as a singular and rich auditory experience. Take a listen to the Beatles' song *Because* for a masterful and inspiring lesson in harmony. All that being said, architecture is not a linear unfolding of events like a musical composition. Architecture operates through all the senses simultaneously, and one could argue architecture, because it is part of an ever-changing natural environment, is also ever-changing and is never the same experience twice.

Architecture

We have discussed the value of the visual, auditory, and tactile in terms of sensation and the capacity of artistic media to awe and inspire us through our senses. Architecture caries these sensual interactions even further. We might smell the dampness of the grotto or basement, the stuffiness of the books in a library, or the aroma of something baking in the kitchen. As we move through the spaces of architecture, we feel the coolness of cold surfaces or the warmth of sunlight passing through a window. We reach out and touch and feel the doorknob, or we may feel the weight of a door as we push or slide it open. We feel the tactility of the floor under our feet and are acutely aware of any slope. We feel the labor of climbing a stair. As we walk, we also hear our footsteps reverberating or echoing in the space, another clue to our active presence and the materiality of the ground. We hear others moving through the space. We see the play of light and shadow projecting two-dimensional contrast that is diffused onto three-dimensional surfaces. We see the changing color of the sky saturating into and being reflected by the surfaces around us as the sun rises, transits, and sets, offering a range of yellow, orange, purple, pink, and blue hues from morning into the evening. We feel our proprioceptive relationship with the spaces around us as we move through them—tall, narrow, large, compact, expansive, or claustrophobic. These "sensations" of spatial size are proprioceptive measures we make based on the proportion of the space to our body. Finally, we take measure of variability of time in the spaces we occupy—seconds as hours or time flying by. As the day light changes, as we initiate and complete activities, and as we may be focused or restless, this temporal sensation is quite dynamic. These aspects of dwelling are a small survey of the opportunities for design to inspire people and suggests how architects might think about how the "technology" we use in architecture is not technological. Our architecture should go far beyond just shaping, organizing, and cooling/heating the spaces people occupy. Architects must conceptualize, design, and animate complete environments.

During his address to architects and educators from around the world in Helsinki, Finland in 2003, speaking in Alvar Aalto's Finlandia Hall, architect and theoretician Juhani Pallasmaa offers this paraphrased quote from Aalto in 1957:

... Architecture has a second thought (...) the idea of creating a Paradise. That is the only purpose of our buildings (...) we wish to build a Paradise on earth for people.

—Alvar Aalto

If we take as an example Pekka Pitkänen's competition-winning Chapel of the Holy Cross near Turku, Finland, built in 1967, we can explore the relation between architecture and the senses in this modest yet powerful reverent architecture of funerary ceremony. Consisting of three chapels (160, 50, and 12 people), people arrive through one entry point and leave the chapel through portals into the landscape leading to the cemetery. The architecture seeks to become part of the environs allowing the landscape and participants to flow through the space while transitioning from the common activities of daily life to those of contemplative introspection of life.

Figure 2.6 Landscape Foyer Entry, Pekka Pitkänen, Chapel of the Holy Cross 1967, Turku, Finland.

The landscape foyer gathers together visitors from multiple directions of arrival as they emerge from different paths through the natural landscape. We feel the roughly hewn footpaths of granite pavers that gently slope up to the entry suggesting deliberative movement ascending to the entry foyer. We feel the breezes and hear the rustling pines in the nearby forest which begins to mask the noises of civilization as they fade in the distance. We smell the coniferous plants as we inhale pinene (pine molecules) which act as a bronchodilator making breathing easier. The architecture offers us an arrival portico, a covered shelter to gather before being invited to cross over a manicured bed of plants as we move from the nature, deeper into the constructed spaces.

Figure 2.7 Portico and Entry Door, Pekka Pitkänen, Chapel of the Holy Cross 1967, Turku, Finland.

Our footsteps echo of the adjacent wall and the ceiling of the portico which also shades us from the sun and sky giving the sense we are "in" a space, yet still "in" the landscape. From the portico, we see plants, paths, and walls in the landscape entering into the chapel foyer suggesting that we do the same.

Figure 2.8 Brass Door Pull, Pekka Pitkänen, Chapel of the Holy Cross 1967, Turku, Finland.

But first, we must touch the architecture. Pitkänen asks us to reach into the door and feel the cold, simple cylinder of the copper door-pull. We see that it is polished against the dark patina of the aged copper door—the touch of those who have come before us. We must pull the heavy door open, feeling its weight and as it opens into our path, we must move with it to negotiate our entry. Perhaps we greet a stranger or loved one with a reverent nod in that moment as we let them enter first.

After we enter the chapel foyer, we might linger there for a moment and recall our walk, or perhaps wait for others to arrive. Here, the concrete walls and ceiling seem to float over us and out into the landscape offering a surface for light to reflect off of and then diffuse into the space while guiding our view outward. We are in the space of the chapel and yet still in the landscape as well.

Figure 2.9 Interior Foyer and Entry Doors to Chapel, Pekka Pitkänen, Chapel of the Holy Cross 1967, Turku, Finland.

Returning to our course toward the chapel, we see that the architect is illuminating the passageway with daylight when available but becoming more judicious with the light allowing our eyes to adapt to the darker chapel space ahead. We can imagine that the beam of diffuse light we see in the image below might be sharp and crisp on a bright sunny day and may take on the color of morning light or the amber purple of a sunset. Set against the light grey concrete, we see the subtleties of this quality of daylight. The passageway narrows relative to the foyer, and auditory, visual, and proprioceptive senses relay this perception to our brain. The opaque copper doors "float" in the space of the passage allowing us to see the space extend beyond them while maintaining privacy between the foyer and chapel space. Again, the weight of the doors is palpable visually and as we pull the doors open, the creaking is amplified as the sound is reflected off of the acoustically hard concrete and glass surfaces in this compact space. Do we pause and hold the doors to close gently, or do we leave the doors to close as they may, with a reverberant thud announcing our presence? We are participants in this choreography.

Figure 2.10 Entering the Chapel Along the Boundary Wall, Pekka Pitkänen, Chapel of the Holy Cross 1967, Turku, Finland.

Making our way down the passageway, we come to the chapel, a space of pause, contemplation, and reflection on the journey to lay our loved one or colleague to rest. We enter the chapel, along the wall with daylight peaking over and guiding us. There is more daylight across the space, suggesting the next transition on the itinerary. The chapel, however, is somewhat dim. Pitkänen uses monochrome concrete to allow the light to diffuse softly in the space. He presents the concrete in its simplest unfinished form offering a velvety exchange with the diffuse daylight that rakes across its surface. We move from light, to light, but for a while, we mourn in the shadows.

In the chapel, wooden benches, much like the ceiling and balcony, float in the space. We touch the wood, not the concrete. We may be here in winter, with heavy clothes or in the summer lightly dressed being more or less insulated from the cool or cold benches. The wood will not absorb our heat as does the concrete with a much higher thermal capacity. We think about the warmth of the wood in contrast to the cold of the concrete.

Behind the organ balcony, skylight is invited to enter in from above as the balcony is pulled away from the wall letting the light fall through and into the chapel space below. We perceive this softening of light as it saturates the surfaces revealing texture and depth. This is of course, one moment in time, but we know this light will run its course throughout the day and the year. While occupying the space, this diurnal change may be imperceptible, yet if a cloud drifts by on a sunny day, it will surely be sensed. The hard, bare, and massive concrete material, coupled with the large volume of the space enhances the power and strength of the organ while the sounds reverberate throughout and persist in the space. Participants are immersed in the aural experience as the music is filling the space and arriving at the listener from all directions. We feel the tactility of the bass sounds from the pipe organ resonating in the wooden benches.

Figure 2.11 Funerary Chapel, Pekka Pitkänen, Chapel of the Holy Cross 1967, Turku, Finland.

Figure 2.12 Skylight and Organ Balcony, Pekka Pitkänen, Chapel of the Holy Cross 1967, Turku, Finland.

Figure 2.13 Skylight Illuminating Altar and Casket Lift, Pekka Pitkänen, Chapel of the Holy Cross 1967, Turku, Finland.

Figure 2.14 Altar, Pulpit, and Path to Cemetery, Pekka Pitkänen, Chapel of the Holy Cross 1967, Turku, Finland.

Figure 2.15 Departure From Chapel to Burial, Pekka Pitkänen, Chapel of the Holy Cross 1967, Turku, Finland.

At the altar, sky light is again employed to define the focal places in the chapel and to convey a sense of reverence within the ceremony. The coffin is actually mechanically elevated into the space from below and into the light for the ceremony to begin.

Within the chapel, we literally and figuratively turn a corner—things are different after the ceremony. On the left from where the benches are facing, we are again offered the view of nature, the idea of returning to the light and to the earth. The entire lower area of the chapel is visually open drawing in the flow of daylight and allowing view to flow out. During the ceremony, family and friends are reacquainted with the departed, brought together in chapel, and prepared to pilgrimage, now as a collective.

As we depart the chapel returning to the natural environment, the architecture deconstructs into the landscape as we pass by trading walls for shrubs, the ceiling for sky, stillness for breeze, and the organ for breezes in the trees. Walking the down the gently sloping path, our stride is less burdened. Perhaps we are less burdened in our mourning and in our loss. We have performed the ritual of recognizing a life well lived.

Pitkänen has given us an architecture of sensation, that like good poetry, connects us to the earth and reveals those things that are meaningful. We are likely not even consciously aware of the brilliance or power of this level of architecture, but we can feel it.

Sensation and Cognition

Figure 2.16 Allegory of the Five Senses, Pietro Paolini, Circa 1630.

We have been speaking of architecture, sensation, and reverence in the Chapel of the Holy Cross by Pekka Pitkänen, a funerary chapel that requires us to consider introspection and thoughtfulness as designers. Pietro Paolini's *Allegory of the Five Senses*, paints a scene of somber elegance by characterizing the more common pursuits of making music, drinking wine, or gambling. By calling into consideration our shared senses, using the technique of Caravaggio's chiaro scuro (light-dark) style, it is suggested that the context of our senses and interpretation is essential to our humanity.

Phenomenalism

John Stuart Mill was one of the first philosophers to argue, as utilitarianism, that we know our world through objects and the influence those objects have on us as perceived through our senses. If we cannot sense something, we cannot know it. Immanuel Kant extends that idea to argue that "we may be able to infer what lies beyond perception, but knowledge is limited to the experienced world," which is sensed. Phenomenalism is not to be confused with Phenomenology which is the study of experience and consciousness, yet the two perspectives are necessarily related. The philosopher Gaston Bachelard, would argue that because we gather experiences through memory, our consciousness experience is more through judgments, perceptions, and emotions and that generally, we cannot perceive things objectively—our world of perception is inherently subjective and being derived, perhaps as much by what we sense as who we are. Our individual experience is interpreted.

We should consider that our interactions with our environment, built or natural, are both sensory and phenomenal experiences. Who we are, comprises much of how we conceptualize the sensations of our environment. The image of the oak in the mangrove is intentionally vague. What does it recall for you in the context of this reading?

Figure 2.17 Oak in Mangrove. The Forms and Lines of Growth, Reach Up and Twisting Down, Are Both Amorphous Yet They Convey a Feeling, a Mood, or Perhaps a Memory That Might Connect You to Another Place and Time, Perhaps Your Childhood.

Psychophysics

As we consider that the world of our perception comprises a subjective reality, we are obliged to understand the basis of that subjectivity by studying the core relationships and systems involved in sensing the world in which we dwell. Science has revealed, so far, at least six factors affecting our perception through stimulus and sensation. Others have been postulated and even those in this list are still questioned to some degree, so even the factors of sensation, are not fully and clearly defined with broad scientific agreement.

Six Factors of Stimulus and Sensation

1. Environment
2. Human Anatomy
3. Metabolism
4. Human Physiology
5. Adaptation
6. Interpretation

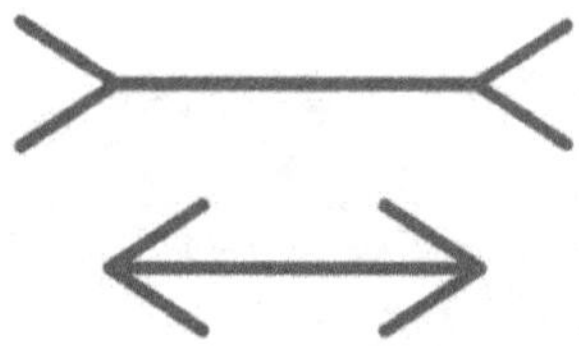

Figure 2.18 Müller-Lyer Optical Illusion—Are the Horizontal Lines the Same Length?.

Let's take a moment here to mention the situation of the illusion—something that appears to be perhaps one thing, but is in fact much different than it "appears" to be. This could be an optical illusion such as the case with the famous Müller-Lyer optical illusion illustrated in the image below. Even though we know the horizontal lines are the same length, the upper line still "looks" longer than the lower line.

Furthermore, we can also experience cognitive illusions that lead us to errors in judgment that are fundamentally illogical. Amos Tversky and Daniel Kahneman—Nobel Prize winning researchers and authors—study cognitive bias and decision psychology. They developed "The Linda Problem," as an example to tease out cognitive illusions that people experience.

Linda is 31 years old, single, outspoken, and very bright. She majored in philosophy. As a student, she was deeply concerned with issues of discrimination and social justice, and also participated in antinuclear demonstrations.

Which is more probable?

a. *Linda is a bank teller.*
b. *Linda is a bank teller and is active in the feminist movement.*

Figure 2.19 Man Mo Temple, Hong Kong. The Air Is Tangible as Incense Vapors Linger Between Candlelight and the Daylight Entering From Above; Water Remains on the Floor From a Recent Rain Falling Through the Open Skylight and Trickling Down the Steps.

Answer a: Yes! This is the most likely scenario as there are many more bank tellers than there are bank tellers who are active in the feminist movement. Only 15% of the people surveyed chose this answer.

Answer b: Sorry, this is a "Cognitive Illusion." Logically, any single condition such as bank teller, is more likely than the conjunction of two conditions. Linda is a bank teller is much more likely. 85% of survey respondents however chose this incorrect answer.

1. Environment

In a previous module, we introduced the definition of the environment as: *the totality of that which surrounds us.* We sense the stimuli of our "surroundings" as heat, cold, humidity, sound, light, air, spatiality, and temporal change. Architects must consider not only the material and structure of a building but how the design of space acts on inhabitants through this array of environmental stimuli.

2. Human Anatomy

The six factors of stimuli noted above are sensed through the sensory receptors of our bodies. Our range of sensation is limited and operates in a complex fashion with regard to the stimulus based on the anatomy of these sense organs: eyes, ears, nose, tongue, skin, and their related anatomical systems—visual, auditory, olfactory, gustatory, and tactile. Our sensitivity to the environment for vision and hearing, for example, is on a logarithmic scale which results in a just perceivable difference even when the stimulus measure is doubled (twice as much). For example, if you want to make your home stereo twice as loud, you might consider doubling the power from, say 100 watts to 200 watts. But that doubling of power would only be perceived as a slight increase—noticeable but certainly not twice as loud. To make your home stereo sound "twice as loud," you would need ten times the power or 1,000 watts if you started with a 100-watt system.

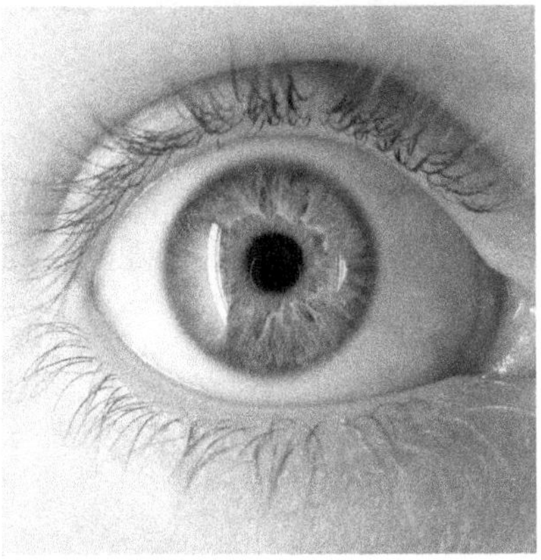

Figure 2.20 Photo of the Human Eye; the Structure and Form of the Eye Shape the Frame of View in We "See" Visual Information.

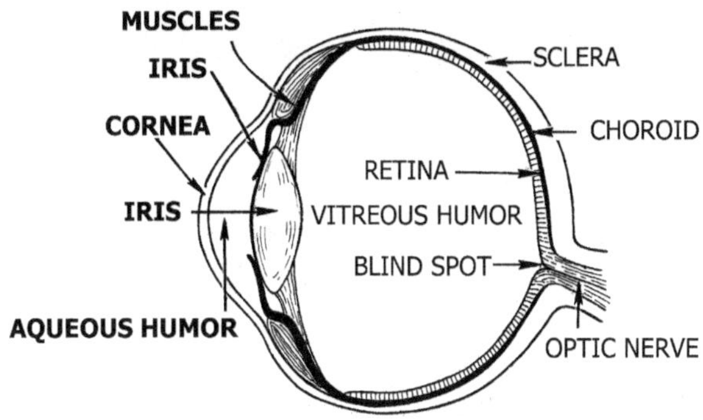

Figure 2.21 Section Diagram of the Human Eye; the Physiology of the Eye Focuses Light, Inverts the Image, and Translates Electromagnetic Waves Into Electrical Impulses That Travel to the Brain. Interestingly, the Brain Inverts the Image so We See Things Right-Side-Up.

3. Metabolism

This is the process of our body converting food into energy in order to do the things we would like to do and to maintain homeostasis—a relatively constant internal temperature of approximately 98.6°F. Human heat production is related to conscious and unconscious activity. As we increase our activity, excess heat is generated and we have to pass that heat to our environment or we will overheat. Physiological reactions to environmental stresses or our physical activity include: perspiration, shivering, expanded or constricted blood flow to limbs, slowing of heart rate, reduction in cognitive capacity, and in extreme cases, shutting-down of nonessential functions.

As people age, their metabolism tends to slow, and may tend to measure physical exertion more acutely. Tom Petty, in his days with The Traveling Wilburys recounts that when promotors were seeking to get the Wilubrys to tour, by telling them how little work they will have to do, and how easy it will be, John Lennon responded: "will it be more work than lyin' in a hammock?"

Figure 2.22 Bodensee Marathon, Germany, 2006.

Figure 2.23 Man Sleeping in a Hammock.

4. Physiology

This is the process by which sensory information is transferred from the sensing organ to the brain. This transmission of information takes time as the stimulus is converted to chemical electrical signals in the body, transmitted through the spinal cord and medulla oblongata to the brain. At the medulla, the left and right side of the nervous system crosses over (left brain controls right side of the body and vice versa) and many autonomous functions such as breathing and heart rate are controlled unconsciously as sensory stimuli are processed. We have often responded to environmental stimuli physically before we are even aware of the stimuli.

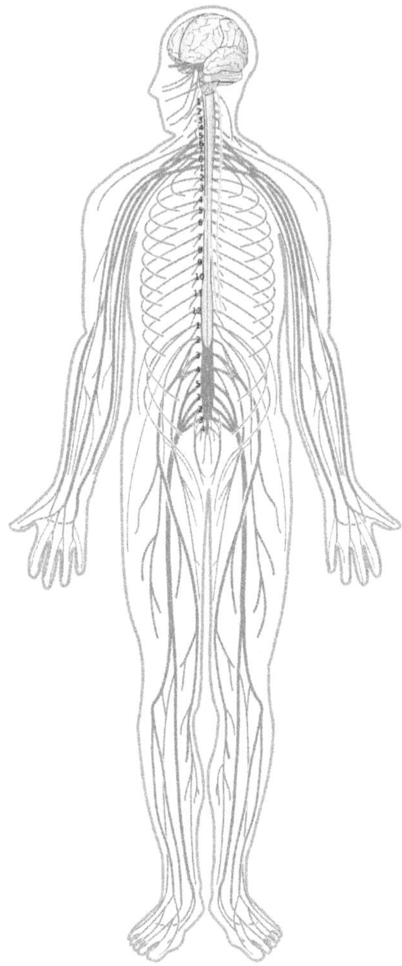

Figure 2.24 Central Nervous System.

5. Adaptation

The human body is quite adept at adapting to acute environmental circumstances over a range of time scales. Visual adaptation, for example, can occur in as short as 5 min if one steps into bright light (photopic) or up to 30 min moving into dark places (scotopic vision) and this is an important issue for people moving into and out of buildings. On the other hand, it can take days or weeks for people to adapt to seasonal changes—or traveling to a different climate—also referred to as acclimatization. In both of these cases, physiological changes occur in the sensing organs.

6. Interpretation

What does all of this sensation that is being transmitted to the brain mean? The influences of past experience and memory (who we are) in addition to what we are or were doing just before the "new" stimulus, and what we may have been thinking about all contribute to our interpretation. Our perspective and interpretation of a stimuli is quite relative to our past (immediate and long term) and our state of mind at the time of the judgement. The famous Rorschach ink blot psychoanalytical tools were used to gain insights into a person's state of mind. What respondents saw in the ink blots (humans, animals, skins, rugs, etc.) says more about the respondents than the blots actually infer—they are just blots of ink.

Temporal change is another interpretive aspect of human experience that manifests in boredom or overtaxing. "Variety is not the spice of life - it is the very stuff of it...," writes suspected spy Christopher Burney about his 526 days in solitary confinement in a German prison outside Paris in 1942. Woodburn Heron, writes of pilots experiencing hallucinations from long monotonous flights. On the other hand, if our tasks are too challenging, and the environmental conditions are not supportive, or the changes in it are too extreme or rapid, we can become overtaxed, fatigued, or even collapse. Our environments must be dynamic within moderate ranges.

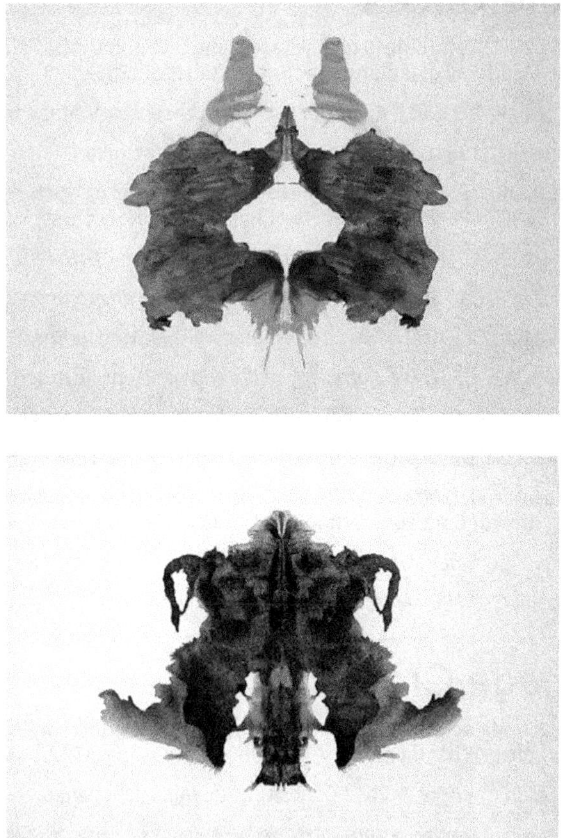

Figure 2.25 Inkblots of the Rorschach Test Printed in Rorschach's Rorschach Test—Psychodiagnostic Plates 02 (Top) and 04 (Bottom).

Summary of Human-Sensory and Environmental Interaction

Perhaps this module has given you some insight into the labyrinthine complexity of relations between human beings and their environments. Sensation itself is quite complex because it involves multitudes of physiological systems that translate the physicality of our surroundings into cognitive assessments and those interpretations draw from our recent and past memory, our state of mind, and our current metabolic condition. It is not fruitful to try and grasp all of these concepts and move forward with the notion that they might be mastered. Rather, it is better to be aware of and perhaps familiar with them as opportunities for architecture to connect with people through a range of sensations and experiences that goes far beyond and is inherently much richer than the merely visual perspective that can be readily rendered in process drawings and models.

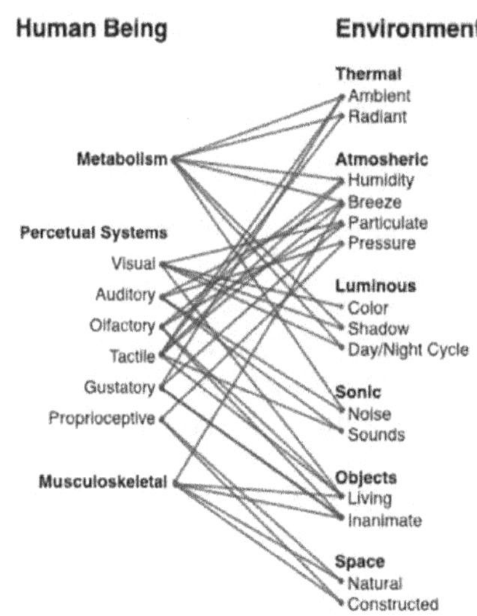

Figure 2.26 Summary of Human Being and Environmental Interactions, Adapted by Martin Gold From James Marston Fitch.

References

Aalto, A. (1972). Taide ja tekniikka [Art and technology] (J. Pallasmaa, Trans.) In G. Schildt (Ed.), *Luonnoksia: Alvar Aalto* (pp. 87–88.). Academy of Finland. (Lecture, October 3, 1955).

Ackerman, D. (1991). *A natural history of the senses*. Vintage Books.

Burney, C. (1984). *Solitary confinement*. Pan Macmillan.

Castellanos, R. (2013). Pekka Pitkänen: The Chapel of The Holy Cross, Turku, 1967, "Pekka Pitkänen: Capilla de la Santa Cruz, Turku, 1967 / Pekka Pitkänen: the Chapel of the Holy Cross, Turku, 1967," with Débora Domingo Calabuig. En Blanco 11, 37–39.

DiLaura, D. L., Houser, K. W., Mistrick, R. G., & Steffy, G. R. (2011). *The lighting handbook* (10th ed.). Illuminating Engineering Society.

Fitch, J. M. (1975). *American building, vol. 2: The environmental forces that shape it*. Schocken Books.

Heidegger, M. (1971). Poetically, man dwells (A. Hofstadter, Trans.). *Poetry, language, thought* (pp. 213–229). Harper and Row.

Heron, W. (1956). *The Pathology of Boredom*. Scientific American.

Kahn, L. I. (1991). Form and design. In A. Latour (Ed.), *Louis I., Kahn: Writings, lectures, interviews* (pp. 116). Rizzoli.

Kahneman, D. (October 1, 2011). *Thinking, Fast and Slow* (1st ed.). Farrar, Straus and Giroux.

Pallasmaa, J. (2003, July 27–29). *Contribution and confusion: Architecture and the influence of other fields of inquiry*. ACSA (Association of Collegiate Schools of Architecture) Conference, Helsinki.

Petty, T. *Commentary on the traveling Wilburys*. Tom Petty Radio, Sirius XM.

Siebein, G. (1986). *Man technology and environment* (2nd ed.). Storter Press.

Image Credits

Fig. 2.1: Michelangelo Merisi da Caravaggio, "The Incredulity of St. Thomas," https://commons.wikimedia.org/wiki/File:Caravaggio_-_The_Incredulity_of_Saint_Thomas.jpg, 1602.

Fig. 2.2: Copyright © 2012 Depositphotos/Alex_Ishchenko.

Fig. 2.3: Claude Monet, "Monet Haystacks, End of Summer.," https://commons.wikimedia.org/wiki/File:Claude_Monet_-_Haystacks,_end_of_Summer_-_Google_Art_Project.jpg, 1891.

Fig. 2.4: Gian Lorenzo Bernini, "David, Gian, Lorenzo Bernini, 1623-1624.," https://commons.wikimedia.org/wiki/File:Bernini%27s_David.jpg, 1624.

Fig. 2.5: Copyright © 2014 Depositphotos/ne_v_focuse.

Fig. 2.16: Pietro Paolini, "Allegory of The Five Senses," https://art.thewalters.org/detail/15244/allegory-of-the-five-senses/, 1630.

Fig. 2.18: Franz Carl Müller-Lyer, "Müller-Lyer optical illusion," Optische urteilstäuschungen. Archiv für Anatomie und Physiologie, Physiologische Abteilung, 1889.

Fig. 2.20: Copyright © 2023 by Martin Gold (CC BY-SA 3.0) at https://commons.wikimedia.org/wiki/File:A_blue_eye.jpg.

Fig. 2.21: Pearson Scott Foresman, "Section diagram of the human eye," https://commons.wikimedia.org/wiki/File:Eye_-_Human_(PSF).png, 2019.

Fig. 2.22: Adapted from Ankawü, "Bodensee Marathon, Germany, 2006," https://commons.wikimedia.org/wiki/File:Start_des_Bodensee-Marathons_2006.jpg, 2006.

Fig. 2.23: Copyright © 2018 Pexels/Frank Meriño.

Fig. 2.24: Copyright © 2014 by Medium69, Jmarchn (CC BY-SA 4.0) at https://commons.wikimedia.org/wiki/File:Nervous_system_diagram_unlabeled.svg.

Fig. 2.25a: Hermann Rorschach, "Rorschach Test Psychodiagnostic Plate 02," https://commons.wikimedia.org/wiki/File:Inkblot_2.tif, 1927.

Fig. 2.25b: Hermann Rorschach, "Rorschach Test Psychodiagnostic Plate 04," https://commons.wikimedia.org/wiki/File:Inkblot_4.tif, 1927.

Fig. 2.26: James Marston Fitch, "Human Being and Environmental Interactions," Historic Preservation: Curatorial Management of the Built World. Copyright © 1990 by University of Virginia Press.

Module 3

Sun Climate Weather

Martin Gold

Architecture appears for the first time when the sunlight hits a wall. The sunlight did not know what it was before it hit a wall.

—Louis Kahn

As Louis Kahn expresses the interwoven relationship between architecture and sunlight, we can see from his design for the Salk Institute, that this influence is just one aspect of his regard for understanding, climate, the nature of sunlight at a particular place, the views offered on his site on the coast of the Pacific Ocean, and the potential for the occupants of the institute to thermally moderate their spaces by adjusting operable windows and louvered shades. The Salk researchers are active participants in embracing the natural climate by closing the windows and opening the louvers to allow sunlight to warm the space, or by opening the windows and adjusting louvers to shade the space while allowing breezes in for cooling, or by opening both windows and louvers for cool breezes and the radiant warmth of the sun. When the windows are open, the scents of the pacific waft in, and the sounds of gulls and turns remind the occupants of their place in the natural environment.

Figure 3.1 Louis Kahn's Salk Institute, La Jolla, California.

Returning for a moment to James Marston Fitch's concept of architecture as the third environment (meso), and the proposition that technology is the interface between this architectural (meso) environment and the natural (macro) environment, we must gain an understanding of the opportunities, structures, and ecologies (systems of structures) in both of these architectural and natural environments. We have discussed human sensation and metabolism (micro environment), and in a module on the topic of thermal comfort, the interface between the micro and meso environments is explored. In this module, we focus on basic principles and systems within the natural environment that are related to our position on the earth relative to the sun and geographical impact it has on climate. This focus will include the influence of local context that can generate microclimates akin to, or in contradiction with, the general climate. As we gain understanding of the resources within these natural environments, we can make informed choices regarding the technologies we deploy toward an architecture that is inspiring, ecologically sensitive, and connects us to our place.

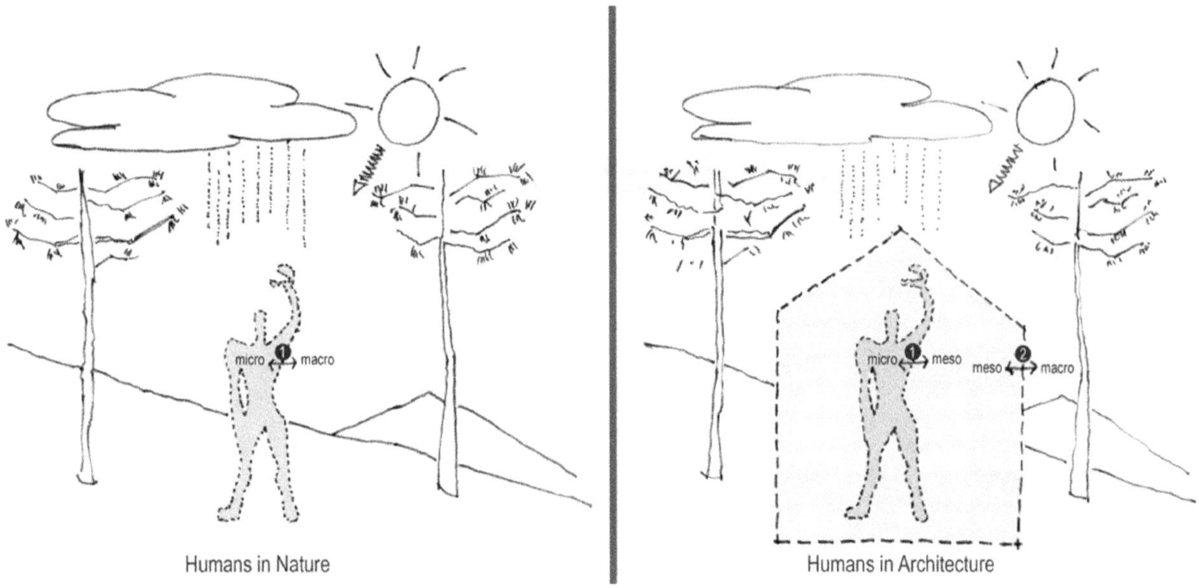

Figure 3.2 Third Environment—Macro (Natural), Meso (Architectural), and Mico (Biological) Environments With Interface 1 of Skin and Clothes and Interface 2 Technology.

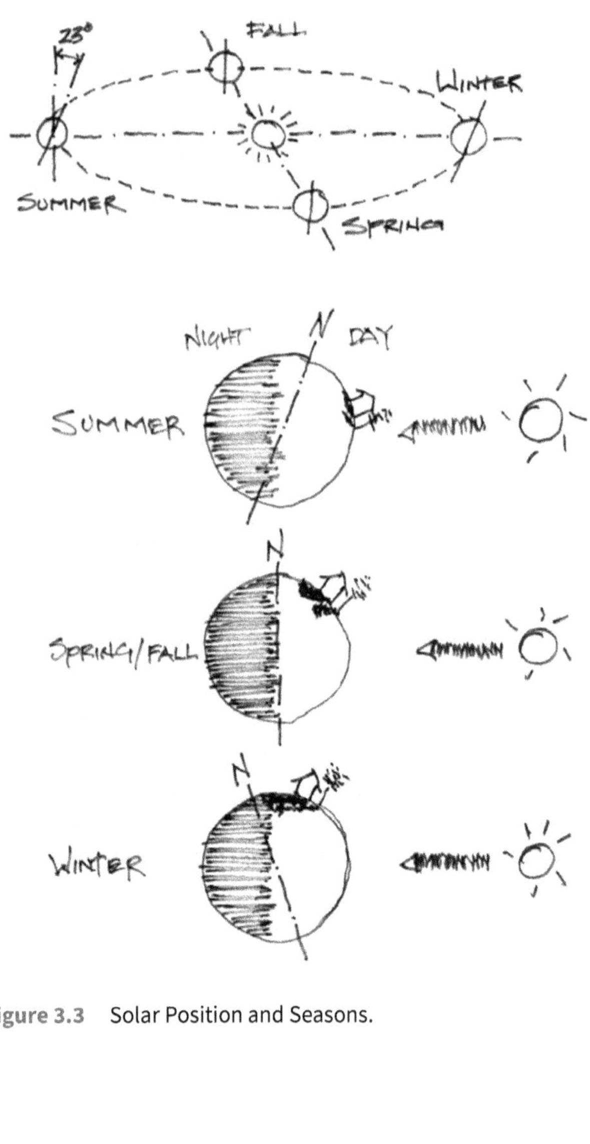

Figure 3.3 Solar Position and Seasons.

Seasons

As the earth completes its orbit around the sun each year, the tilt of the earth's axis, at 23°27' from the orbital plane, gives us our seasons. In summer (northern hemisphere), the tilt is toward the sun, and the sun's rays arrive more perpendicular to us when we move away from the equator. In winter, it is just the opposite, we are tilted away from the sun, and the sun's rays arrive at more of an angle and subsequently are less intense. During fall and spring, we transition between the solstices (summer and winter), passing through the equinoxes (fall and spring) when both day and night are the same length. Both the northern and southern hemispheres have the same seasonal changes. Over the year, this dynamic heating and cooling of the earth's poles, and the difference in heating between the equator and the poles, drives much of our weather patterns.

Climate Zones

As the earth orbits the sun, tilting both toward and away during a given year, coupled with the Earth's spherical form, the sun impacts the earth differently as one moves toward the poles from the equator creating identifiable climate zones:

Equatorial Zone

Here the sun directly overhead twice during the year as it tracks from south to north and back again across the equator. The equatorial zone gets both perpendicular and near perpendicular solar radiation throughout the year and is closer to the sun than the rest of the planet. Most of the solar radiation in this zone strikes buildings on the roof with much less striking the building's walls. Here, even small overhangs can provide ample shade on buildings. Days and nights in this zone have equivalent durations throughout the year with relatively short dawn and dusk periods.

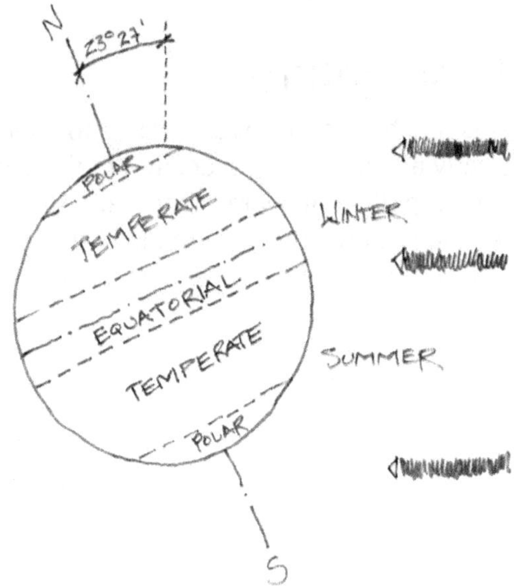

Figure 3.4 Basic Climate Zones Relative to Annual Solar Exposure.

Temperate Zones

These occur in both the northern and southern hemispheres and are characterized by more angular solar radiation than the equatorial zone and being somewhat more distant from the sun which reduces the solar intensity. Here, the sun can strike buildings on both the walls and the roof. The ratio from striking more on the roof, to striking more on the walls increases as one moves toward the poles. As one moves farther north in this temperate zone, the duration of the day to night also changes significantly—longer days in summer and longer nights in winter.

Polar Zones

These zones receive solar radiation that spreads across a much larger area of the earth due to the grazing angle of solar energy thus reducing the sun's intensity. Here, the sun predominantly strikes the walls and windows of buildings with minimal exposure to the roof surface. Even in midday in the summer, the sun is very low in the sky. In winter, if we are above the arctic circle, we will not see the sun directly but we will see a luminous sky as the sun is just over the horizon, even at midday. In the summer, the sun returns to the sky and may be shining for 24 hours but it will always be very low in the sky. In the polar summer, the sun strikes all the sides of our buildings (N, S, E, W) as it transits from north through the southern sky and back to north again.

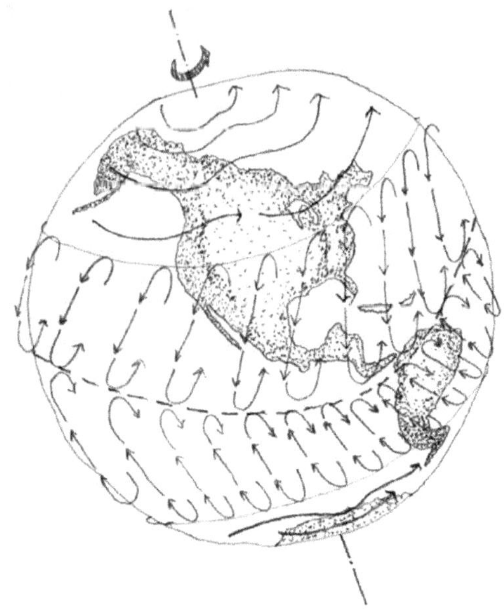

Figure 3.5 Global Convective Forces and the Coriolis Effect Drives Global Weather Patterns.

Global Weather Patterns

The north and south poles receive much less annual solar radiation becoming quite cold in winter, while the equatorial zone receives much more annual solar radiation. Solar radiant heat warms the earth and subsequently, the earth warms the air near the surface. This earth/air temperature difference between the equatorial zone and the polar zones and the high-altitude air (cooler typically) temperature differentials, generates global convective loops within the atmosphere. Furthermore, the earth's surface is rotating faster at the equator (approximately 1,000 mph) than it is at the poles (actually, 0 mph at the polar axis). Therefore, as the air moves toward the poles and then back toward the equator, it is deflected because the earth is basically moving underneath it—faster at the equator and slower at the poles. This force, along with differential solar warming, drives the dominant weather patterns on earth.

Temperature

Seasonal temperatures are related to the basic equatorial, temperate, and polar climate zones but are also greatly affected by altitude, mountain ranges, oceans, and other large bodies of water. Land masses heat more rapidly than water bodies and these local temperature differences create regional or localized convection patterns that in turn affect the local temperatures. In figures 3.6 and 3.7 below, compare summer and winter temperatures with respect to the basic climate zones and continental land masses.

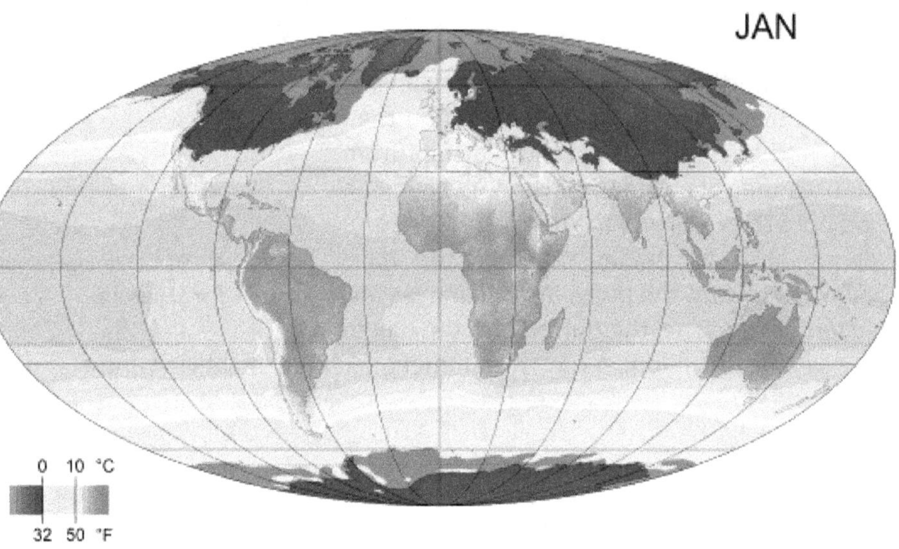

Figure 3.6 January Mean Temperatures From 1960–1990 (NASA).

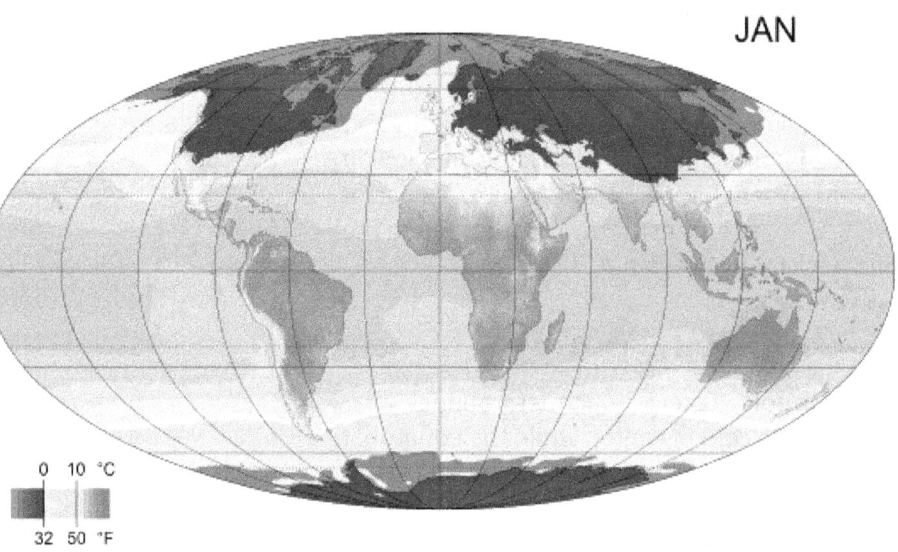

Figure 3.7 July Mean Temperatures From 1960–1990 (NASA).

For an animation of mean global temperatures from 1880–2020, visit the NASA site: https://svs.gsfc.nasa.gov/4964

In the early 2020s, it is generally accepted that global temperatures are rising, and the cause is both carbon dioxide and methane being released into the atmosphere. The chart below shows annual mean temperatures from 1880 to 2020. The data is normalized using the average temperatures from 1951–1980 as baseline reference. Since the year 1880, temperatures have increased 1.1°C (2.0°F). Just a 1°C (1.7°F) drop caused the "Little Ice Age" a fairly recent period of mountain glacial expansion from the 16th to mid-19th century during which the European continent climate was strongly impacted. A 5°C (9°F) drop in average annual temperature, blanketed North America under towering masses of ice that were 2,000 feet thick over Manhattan.

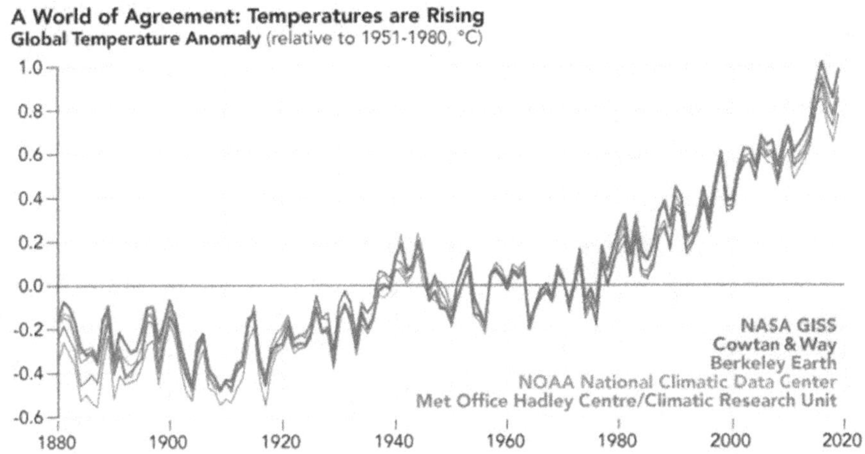

Figure 3.8 Global Temperature Rise 1880–2020 (NASA) for an Animation of Mean Global Temperatures From 1880–2020,

Visit the NASA Site:

https://svs.gsfc.nasa.gov/4964

One of the main contributors to climate change (also referred to as global warming), is the addition of the so-called greenhouse gas carbon dioxide (CO_2). Sunlight, which has a high proportion of ultraviolet (short wave) light, passes through the carbon dioxide in the atmosphere and is absorbed in the earth as heat. Earth's surfaces then re-radiate the heat as infra-red (long wave) radiation. However, that long wave radiation cannot pass back through the carbon dioxide gas in the atmosphere and the heat is subsequently trapped and remains in the atmosphere. With lower concentrations of carbon dioxide, more heat would radiate back out into space. The glass that we use in buildings and cars acts much the same way as carbon dioxide in terms of trapping re-radiated infra-red heat that can be quite useful in maintaining greenhouses in cold environments or warming our homes in winter.

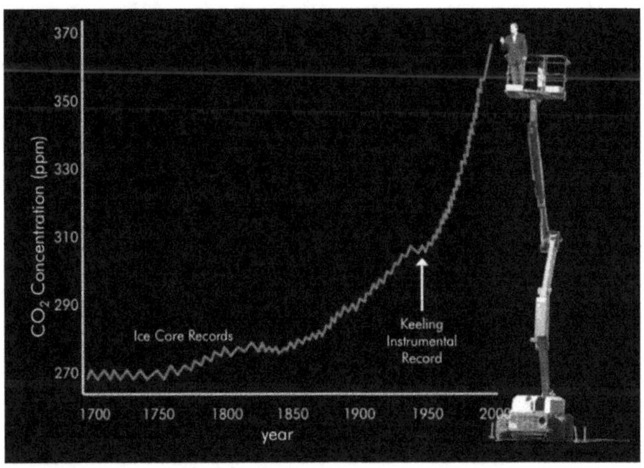

Figure 3.9 Tracking Global Atmospheric Carbon Dioxide Levels 1700–2000; Former U.S. Vice President Al Gore Is Shown Point Out the Alarmingly High Levels Relative to the Historical Record; Interestingly, the Industrial Revolution, in Which Western Cultures Transitioned From Farming and Hand-Making to Machine Manufacturing, Began in 1760 in Brittan.

Clouds

Cloudiness is a visual indicator of the amount of moisture in the air and depending on the location, can range greatly from the pacific northwest, that is cloudy 70%–80% of the year to the southwest U.S. or north Africa which is cloudy less than 30% of the year. The image below is the global 5-year average cloud cover for the month of June. One can see the locations globally where there are deserts. At the link provided below, you can view NASA's animation of global cloud cover and see areas that are predominantly cloudy or predominantly clear over time. Some areas, like the Midwest U.S., are cloudy for long periods of time while others, such as the Southeast U.S., are typically clear in the morning yet get cloudy in the afternoon and evening. Clouds can keep a region cooler as they reflect and absorb arriving sunlight. At night, on the other hand, the clouds reduce heat from radiating to the cold night sky. Understanding the nature of cloudiness is critical in terms of assessing the potential for passive architectural systems and the potential for rainwater harvesting. Clouds of course, are also the indicator of potential rainfall.

Figure 3.10 Global Mean Cloudiness for the Month of June. One Can Find the World's Deserts as Those With Little Cloud Cover.

Link to NASA Cloudiness Animation:

https://earthobservatory.nasa.gov/global-maps/MODAL2_M_CLD_FR

The Mean Sky Cover map below shows cloudiness during daylight hours in the U.S. We might gain insights from this information in terms of evaluating rainwater harvesting, photovoltaic electricity generation, passive heating strategies, and even building materials in terms of retaining or shedding heat quickly. Even in areas with 50% cloud cover, photovoltaic can be quite effective and that 50% cloudiness typically suggests rainwater harvesting is potentially viable. Cloud cover also suggests less extreme daily temperature fluctuations as the clouds limit solar exposure during the day and limit heat loss at night. On the other hand, areas in yellow, orange, and red on the map, have high diurnal temperature swings—hot days and cold nights—that suggest high mass materials could be utilized to mitigate the dramatic temperature differences.

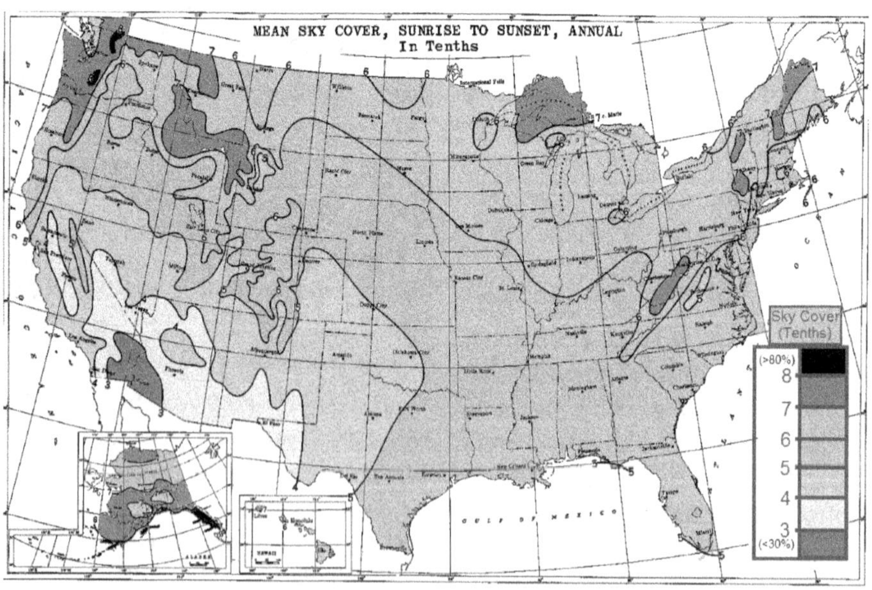

Figure 3.11 Mean Sky Cover in Tenths (Percentage of the Daylight Hours); This Map Is a Helpful Tool for Estimating the Viability of Rainwater Harvesting or the Potential Effectiveness of Photovoltaic Electric Power Generation.

Climate Regions

Within the three general climate zones introduced earlier, there are five more detailed climate categories including: Tropical (A), Dry (B), Temperate (C), Continental (D), and Polar (E). Within each of those, there are up to seven subclassifications. Originally published by climatologist Wladimir Köppen in 1884, the system was revised in 1918 and again in 1936 and is currently accepted as the best resource for climate regions at this scale. The Köppen system takes into account regional characteristics such as mountains, general elevation, waterbodies, and vegetated areas such as forests, prairies, or swamps. Local site conditions can vary greatly from the general climate typology and should be closely evaluated relative to the Köppen system. In comparing the sky cover and climate region maps, one can find clear correlations. The link below will connect to the global Köppen system and more detail regarding each of the climate subcategories.

Köppen Climate Types of the United States

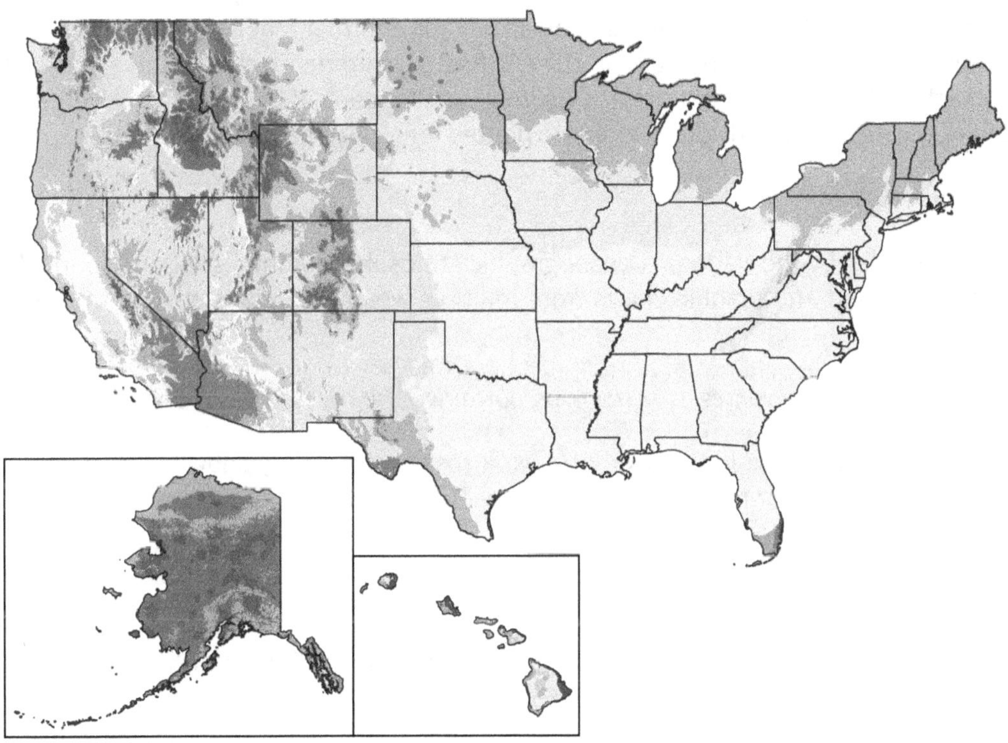

Köppen Climate Type

- Af (Rainforest)
- Am (Monsoon)
- Aw (Savanna)
- BWh (Hot desert)
- BWk (Cold desert)
- BSh (Hot semi-arid)
- BSk (Cold semi-arid)
- Csa (Hot-summer mediterranean)
- Csb (Warm-summer mediterranean)
- Csc (Cold-summer mediterranean)
- Cwa (Humid subtropical)
- Cwb (Subtropical highland)
- Cfa (Humid subtropical)
- Cfb (Oceanic)
- Cfc (Subpolar oceanic)
- Dsa (Hot-summer mediterranean continental)
- Dsb (Warm-summer mediterranean continental)
- Dsc (Dry-summer subarctic)
- Dwa (Hot-summer humid continental)
- Dwb (Warm-summer humid continental)
- Dwc (Dry-winter subarctic)
- Dfa (Hot-summer humid continental)
- Dfb (Warm-summer humid continental)
- Dfc (Subarctic)
- ET (Tundra)
- EF (Ice-cap)

Data sources: Climate normals from PRISM Climate Group, Oregon State University, https://prism.oregonstate.edu; Outline map from US Census Bureau

Data periods: 1991-2020 (Contiguous United States); 1981-2010 (Alaska); 1971-2000 (Hawaii)

Figure 3.12 Köppen Global Climate Classification System.

Link to Detailed Climate Classifications:

https://en.wikipedia.org/wiki/Köppen_climate_classification

Climate Tools

There are multiple climate tools available both online and embedded within commercially available computer design software such as Rhinoceros and Revit. These climate tools allow users to download climate data, typically in EPW (Energy/Plus Weather Format) format that compiles accumulated weather data for each of the 8,750 hours of the year for decades. It is possible to select average temperatures over a given time range such as the last 13 years or from the full accumulated record, or just one year. The weather data is typically collected at designated weather stations around the globe. One must choose the nearest station that is most similar to the context under evaluation or project site. For example, for a coastal site, one might have to choose a more distant coastal weather station as the nearest station may be located significantly inland of the coast and therefore, will have a different micro-climate.

Climate Consultant: https://www.sbse.org/resources/climate-consultant

Stand-alone program that organizes climate data in a graphic format that is easy to evaluate and understand. This application plots hourly data on the ASHRAE psychrometric chart allowing users to choose passive and active thermal control strategies. Estimates of the percent of time that a given strategy (passive or active) will be able to provide comfort are plotted for a given data set. Free to users.

Climate: https://www.climate.gov

Climate data provided by the National Oceanographic Atmospheric Administration. Hourly data can be selected and organized into graphic charts from multiple weather stations around the globe. Free to users.

Wind Rose Data Set: https://www.nrcs.usda.gov/wps/portal/wcc/home/climateSupport/windRoseResources

A catalog of wind/breeze data in the form of a wind rose that typically shows annual wind direction, speed, percentage of the year the wind blows from each direction, and the percentage of calm periods. Free to users.

Rhinoceros Grasshopper Ladybug: computer-aided design software (CAD)—purchase

In recent versions of Rhinoceros (Windows and Mac OS), both Grasshopper and Ladybug tools have been embedded in the software in the "tools" section. Ladybug allows users to download hourly weather data (EPS) files and by using the Grasshopper visual programming tools, to convert that data into graphic representations that are plotted in the drawing area of Rhinoceros.

Sketchup Safaira: computer-aided design software—purchase

Sketchup software has a Safaira plug-in and a more recent but perhaps less robust embedded climate tool called Architosh to suggest early design climate responses. The Safaira plug-in uses climate data to model building performance rather than present the climate data.

Revit Autodesk: computer-aided design software—purchase

Revit accesses the Autodesk Climate Server to allow downloading of EPS files and the formatting of that data into climate summaries, wind roses, and comfort analysis.

These tools represent popular options among many others and depending on what tools users prefer. There are usually options for engaging climate data within these drawing tools and perhaps, later in the deisgn process, using that data for modeling comfort and energy efficiency of a given design. This list is not exclusive and students are encouraged to search for other options as new programs regularly emerge in support of climate-responsive design.

Balance Point Temperature

The simple definition of Balance Point Temperature (BPT) is the outside air temperature at which no heating or cooling is needed in a building. For example, if we had an office building with a lot of people, lights, computers, and servers generating heat, and the sun was shining on the building, imparting heat, we will likely need to cool the building, even when it is cold outside, or it would get too hot inside. However, if the outside temperature is quite low, then we would also be losing heat, through the building envelope, to the outside air. As you are surmising, there is, for each building, under nominal operations, an outside temperature at which the building would lose, just as much heat as it is gaining—the BPT. For many office type buildings in the United States, this is typically about 65°F, and that value is used generally for calculating the severity of cooling and heating seasons. For large floor area office buildings with lighting, people, and equipment, the BPT can be calculated and could be as low as 35–38°F. One might ask, do we really want to insulate a building that is trying to lose heat even when the air temperature is 35°F?

Heating and Cooling Degree Days

Heating and cooling degree days are a metric used to assess the severity of a climate, in terms of the amount of heating or cooling that might be required to maintain a comfortable temperature. This metric utilizes the balance point temperature (BPT) discussed above as a baseline—no cooling or heating is required. Heating and cooling degree days are calculated using the mean temperature for a location for a given day and then subtracting that temperature from the BPT as shown in the equation below. Note, that if the value is zero, or less than zero, it is not considered. Only positive numbers are accumulated for that day.

For example, if the low temperature for the day is 22°F and the high for the day is 38°F, the math would be (22 38)/2 = 30. 30°F is the mean daily temperature. Then using the 65°F BPT, we have 65 - 30 = 35. 35 is the heating degree day (HDD) value for that day. Each of the 365 days is added together for the annual HDD total.

$$65°\text{F} - \frac{\text{DAILY HIGH TEMP} + \text{DAILY LOW TEMP}}{2} = \text{HDD}$$

Figure 3.13 Heating Degree Day Simplified Calculation Method.

Cooling degree days are calculated basically the same way, except the BPT is subtracted from the mean daily temperature. For example, if the high for the day is 97°F and the low for the day is 86°F, then (97 86)/2 = 91.5. 91.5°F is the mean daily temperature. Then using 65°F as the BPT, we have 91.5 - 65 = 26.5. 26.5 is the cooling degree day (CDD) for that day. Just like heating degree days, we add them up for each of the 365 days of the year.

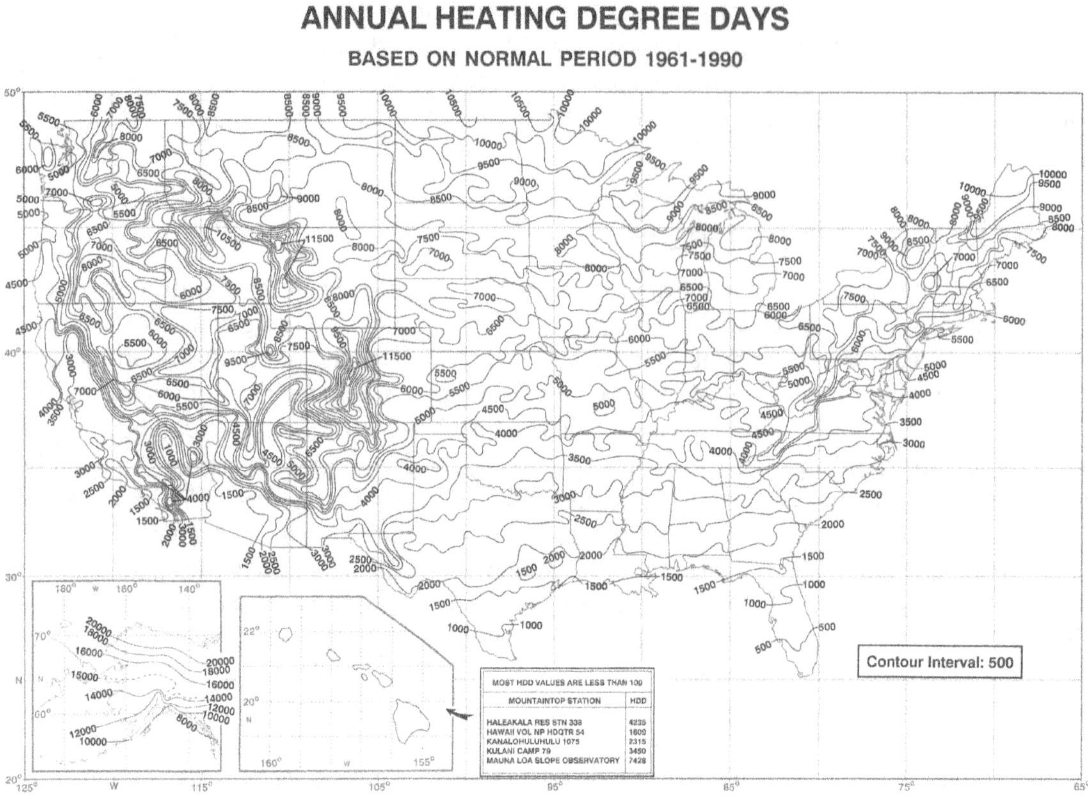

Figure 3.14 Heating Degree Day (HDD) Contours for the United States.

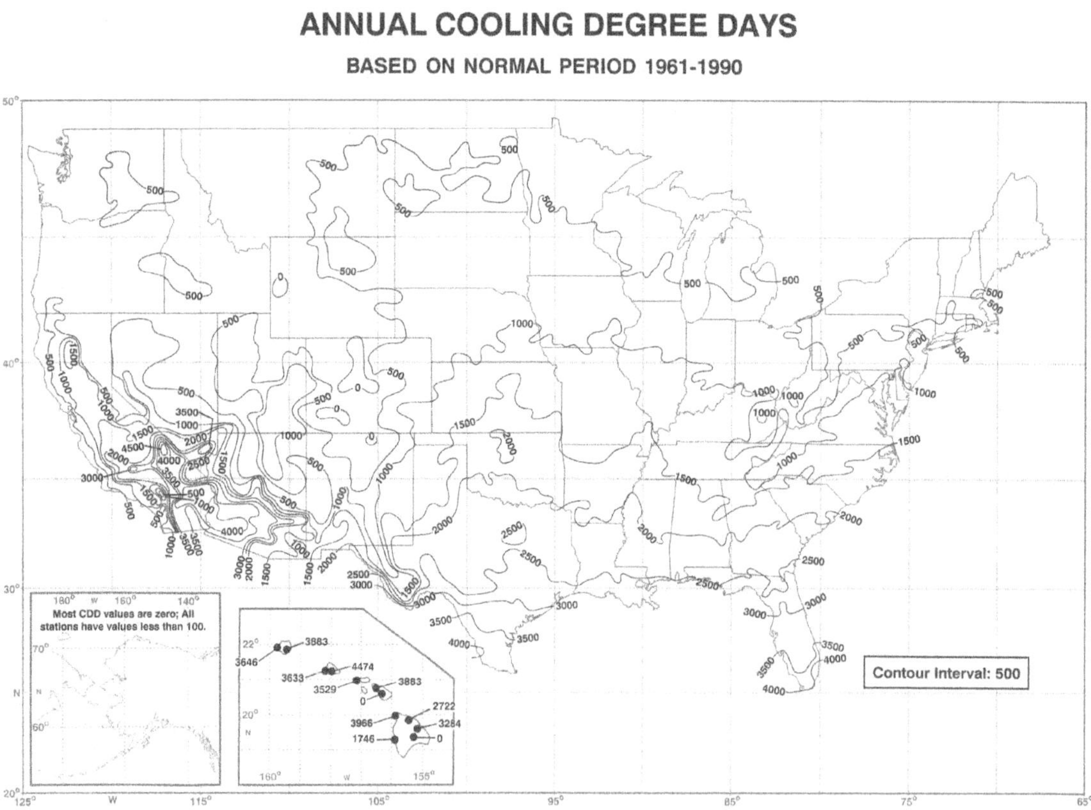

Figure 3.15 Cooling Degree Day (CDD) Contours for the United States.

Heating and cooling degree day contours give us a clear perspective with regard to the severity of heating or cooling required for a particular climate. With that information in mind, we can begin to prioritize active and passive design strategies, consider appropriate materials, and get a sense of the types of mechanical systems that would be appropriate to maintain comfort in buildings.

Location	HDD	CDD
Manhattan, NY	6,170	667
Madison, WI	7,642	467
Phoenix, AZ	1,442	3,746
Miami, FL	199	4,095
Savannah, GA	1026	3549

Figure 3.16 Example HDD and CDD Metrics for Sample U.S. Cities. Notice the Difference Between Clearly Cooling-Dominated Cities (Miami) and Heating-Dominated Cities (Madison); Also, Notice Phoenix and Savannah Are Actually Quite Close in Their HDD and CDD Balance, Yet They Do Have Quite Different Climates When It Comes to Humidity.

Micro Climates

The local context usually offers some unique aspect that generally fits within the overall and sub-climate regions, and the general strategies will prevail. However, understanding how the general climate and weather interact with the site conditions is necessary and will offer insights toward thoughtful and inspiring climate-responsive design.

The coastal estuary of Longboat Key offers a micro-climate this quite cooler than areas just one mile inland from the coastal location. The Key has large water bodies on both sides that moderate temperature locally—cooler when inland areas are hot and warmer when inland areas are cold. The same proximity that captures cool breezes of the Gulf of Mexico or Sarasota Bay, also leaves the architecture exposed to high winds, driving horizontal rain, and wind-blown sand that can deteriorate many common building materials. The low lying area is prone to flooding during storms which is much less likely than even a half-mile inland on higher ground.

Figure 3.17 Longboat Key, Florida, Beach and Estuary Ecology.

Nestled in the clearing of a pine and birch forest near the arctic circle, Alvar Aalto's Villa Mairea is protected from the unceasing winter winds that blow from the northwest. The house is on a promontory that keeps it above cold air masses that might pool in the lower lying agricultural fields. Aalto also wraps the house around a lawn forming an ell shape to protect from the northern winds and opens the courtyard to the southern sun. The clearing is wide to the south to allow the low angled sun of this near polar region to penetrate the interior spaces as the low sun's rays slice over the treetops.

Figure 3.18 Villa Mairea, Noormarkku, Finland, Alvar Aalto, a Promontory in Pine Woods.

The farmhouse and garden in the picturesque town of Vals, Switzerland sit close to the valley on the southeast side of a mountain. In the summer, cool air pools in the valley and as the sun warms the mountain face, the air on the face of the mountain rises, drawing a cool breeze out of the valley. In winter, the cold air pools in the valley, so it is much better to be a bit higher up on the mountain face than at lower areas of the valley. However, if the home is too high up the mountain, then it is exposed to harsh winter winds that graze the upper mountain faces.

Figure 3.19 Farmhouse and Garden, Vals, Switzerland, Near the Valley of the East Facing Mountainside.

In the megalopolis of Manhattan, the local climate is a complex of open river areas providing moderate nominal temperature swings but they also allow for harsh winter winds to impact the city and buildings. Newer buildings, with primarily glass skins, reflect much of the radiant heat from the sun to the adjacent buildings. As the sun transits the sky, this reflected solar heat will be directed to multiple nearby buildings—coined traveling loads. Conversely, the resulting shadows from tall buildings may deprive adjacent buildings of warming sunlight during periods of the day. Within dense urban high-rise cities, microclimates can be very site specific.

Figure 3.20 Hudson Yards, Manhattan, New York, Megalopolis Adjacent to the Hudson River.

Summary

For architects, designers, and constructors, to build more sustainable buildings that seek to connect their inhabitants to the amenities and opportunities in the natural environment, they must become students of the climates in which their buildings, and their inhabitants will dwell. This module offers an overview of the basic principles and physics of the global climate and some detail of the influence of sub-climates within the Köppen climate region map. The discussion of heating and cooling degree days offers one quick estimate of the severity of climates in terms of the magnitude of heating and cooling necessary that will influence mechanical system selection. Many commercially available computer-aided design software programs have climate applications or tools that can download and graphically display detailed climate data for thousands of locations around the world giving designers easy access to this information. There are free applications that organize climate data in illustrative ways that can be meaningful and useful to a designer. In some cases, those tools prioritize and catalog appropriate passive and active design strategies that can be employed to reduce energy use and better connect those dwelling in buildings to their natural climates.

References

Allen, E. (2005). *How buildings work: The natural order of architecture* (3rd ed.), Oxford University Press.

Fitch, J. M. (1975). *American building, vol. 2: The environmental forces that shape it.* Schocken.

Guggenheim, D. G. A. (2006). *An inconvenient truth* [Film], Paramount Pictures.

NASA Earth Observatory. (2023, October 12). *World Of Change: Global Temperatures.* Earthobservatory.nasa.gov. https://earthobservatory.nasa.gov/world-of-change/global-temperature

Image Credits

Fig 3.6: Copyright © by Pzmaps (CC BY-SA 3.0) at https://commons.wikimedia.org/wiki/File:MonthlyMeanT.gif.

Fig 3.7: Copyright © by Pzmaps (CC BY-SA 3.0) at https://commons.wikimedia.org/wiki/File:MonthlyMeanT.gif.

Fig. 3.8: NASA, "Global Temperature Rise 1880-2020 (NASA)," https://earthobservatory.nasa.gov/world-of-change/global-temperatures, 2020.

Fig. 3.9: Lawrence Bender Productions and Participant Productions, "Global Atmospheric Carbon Dioxide Levels," An Inconvenient Truth. Copyright © 2006 by Lawrence Bender Productions and Participant Productions.

Fig. 3.10: NASA, "Global Mean Cloudiness for the Month of June," https://www.nas.nasa.gov/SC21/digitalswag.html, 2021.

Fig. 3.11: NOAA, "Mean Sky Cover in Tenths," https://www.weather.gov/ctp/HowCloudyIsPA, 2022.

Fig 3.12: Copyright © by Icy98 (CC BY-SA 4.0) at https://commons.wikimedia.org/wiki/File:K%C3%B6ppen_Climate_Types_US_50.png.

Fig. 3.14: NOAA, "Heating Degree Day (HDD) Contours," https://commons.wikimedia.org/wiki/File:United_States_Heating_Degree_Day_map,_1961-1990.jpg, 2001.

Fig. 3.15: NOAA, "Cooling Degree Day (CDD) Contours," https://commons.wikimedia.org/wiki/File:United_States_Cooling_Degree_Day_map,_1961-1990.jpg, 2001.

Image Credits

Module 4

Thermal Comfort & Metabolism

Martin Gold

This porridge is much too cold!

Introduction

Human beings can tolerate a wide range of thermal conditions for short periods of time. However, comfort is necessary for our long-term health and well-being. If we are too cold or too hot, we become fixated on the condition, and our body's metabolic functions begin to take control of our energy in both dramatic and subtle ways. If we are too hot, the metabolism is slowed to reduce heat generation, and as a result, cognitive and bodily functions become reduced, resulting in symptoms of dizziness, confusion, headache, cramps, and in extreme cases, we can be rendered unconscious or die within 1–2 hours if we are not cooled down. On the other hand, if we are too cold, the metabolic system once again takes control by reducing blood flow to the body's extremities, activating the muscles to produce heat by shivering and forming goose bumps on the skin. Here too, our cognitive and motor function becomes greatly reduced, and if a person is not warmed and in dire cases, they can succumb to hypothermia. In freezing cold temperatures, hypothermia can happen in minutes to an hour.

Figure 4.1 Thermal Comfort Comes from a Balance of Cool and Warm, Surfaces We Touch, the Air Temperature and Movement Around Us, and Perhaps the Warmth of the Sun.

Architects, of course, are usually concerned with moderate temperatures in controlled environments rather than the extreme conditions of the natural environment. Yet, far before we feel these extreme responses noted above, our metabolism will respond to the subtle drifting of the thermal environment outside of what we perceive as comfort. These conditions can cause minor distractions leading us to change the temperature, put on a sweater, or, more commonly in larger centrally controlled buildings, get a space heater to keep warm. If it is too humid or hot, it is quite difficult to make personal adjustments to our environment, and metabolism and cognition slow down.

Thermal Comfort:

The state of mind when satisfied with the thermal environment.

-ASHRAE

Thermal comfort has been linked to our health, well-being, and productivity.

—Zhang, A. and Frontczak, M. W. P.

Thermal comfort has ranked as one of the highest contributing factors influencing overall human satisfaction in buildings.

—Zhang, A. and Frontczak, M.W.P.

In a 2004 study by Cornell University researcher Alan Hedge, office workers reduced typing errors by 44% and increased productivity by 150% when the ambient temperature was increased from 68°F to 77°F, citing an estimated savings of about $2.00 per hour per employee. Emerging from the late 1970s policies of dramatic energy reductions, setting the wintertime thermostat to 68°F (78°F in summer) became recommended, and in some public buildings, these required set point temperatures. However, if we consider the costs of discomfort, one needs to question the effectiveness of energy saving strategies as part of overall economic production. Is the energy savings effective if people are uncomfortable and it creates errors and reduces productivity? This is especially true for creative culture employees who rely heavily on cognitive focus and ideas as the core generator of their output. Even moderate discomfort can reduce focus, cause fatigue, and contribute to absenteeism. In a study published in the *ASHRAE Journal* in the early 2000s, approximately $320–$1,200 annually per person was lost due to temperature fluctuations in office buildings. With a staff of 100 people, that's almost $8,000 annually on average. There were higher productivity losses for poor layout, excessive noise, and lower losses for the poor lighting.

Architects must nurture a keen understanding of the factors affecting human comfort. This is much more complicated than just setting the Heating Ventilating and Cooling (HVAC) system to a prescribed temperature. How people are dressed as appropriate for the program of the space; what they are doing in the space; radiant solar gain from the sun shining into the space; or very cold surfaces in proximity to people are all factors that are as important as setting a comfortable temperature on the thermostat.

Thermal Comfort

Our body seeks to maintain our internal body temperature of between 97.7°F and 99.5°F and the temperature is lowest in the morning and highest in the afternoon. To do this, the hypothalamus in the core of our brain near our brain stem collects signals from various parts of the body and responds by sending signals back to regulate our metabolism. If we are sedentary and the temperature is comfortable, say 75°F, we should be comfortable. But if we are doing strenuous physical activity or even light physical activity, this temperature may be too high for comfort as we are generating more heat than we can shed at that temperature and may overheat.

> Consider the Minnaert Building, Utrecht, The Netherlands, by Neutelings Reidijk architects. Rainwater is allowed to fall into the building offering a pool of cool water (thermal mass) that also provides evaporative cooling in the summer. However, this may cause too much humidity in the cooler winters.

Thermal comfort—A feeling of well-being or lack of distress

Personal factors:

1. Metabolic state
2. Clothing
3. Acclimatization

Environmental factors:

4. Air temperature
5. Mean radiant temperature
6. Air motion
7. Humidity

If we are losing heat to the environment, our metabolism responds by increasing bodily activity by fidgeting, wrapping our arms together, shivering, and constricting blood flow to the extremities. Restricted blood flow is why our hands, feet, and nose often become the coldest first. If we start to feel uncomfortably cold, our metabolism has likely already responded to the cold environment. If we are gaining too much heat from the environment, we will begin to perspire and may feel sticky or clammy. When moisture on the skin (perspiration) evaporates, it is able to carry away large amounts of heat. If the air around us is very humid and hot, we may not be able to lose enough heat through evaporation, and our clothes will become damp or wet. During this process, the body will begin to try and lower other metabolic functions, making us feel fatigued and lethargic.

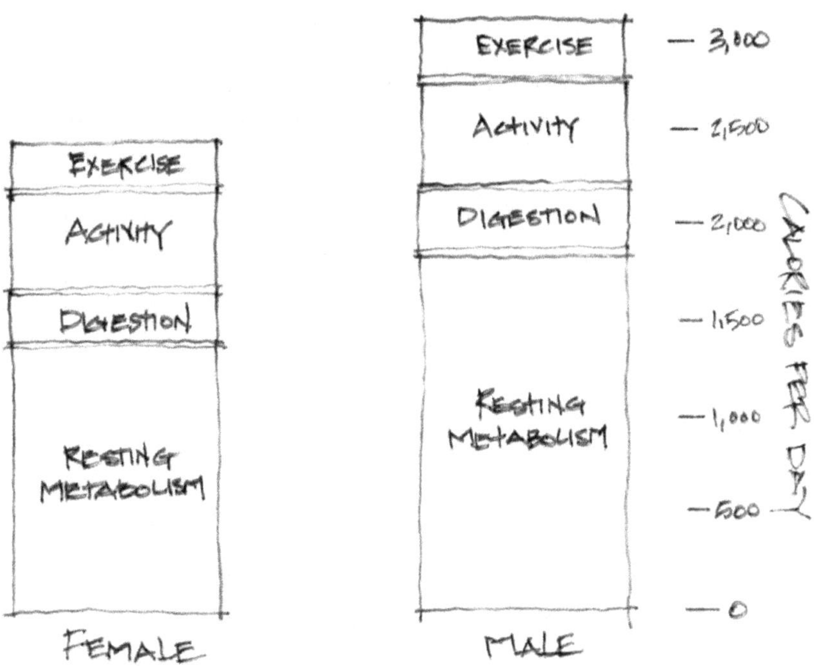

Figure 4.2 Metabolic Rate for Activities by Gender.

Personal Factors Effecting Comfort

Figure 4.3 This Young Girl Enjoys Coloring on a Temperate Day, Wearing Light Clothing, Under Diffused Radiant Heat of the Sun (Slightly Overcast), Pleasant Air Temperature, and Cool Ground Temperature While Doing Sedentary.

Personal factors involve our metabolic state—what activity level we are operating at—and our choice of clothing for that activity. We can become acclimatized to a particular region (hot or cold) or to our dominant occupational condition (indoors or outdoors). In designing spaces that require comfort, we must consider these human aspects of comfort.

Metabolism

A human's chemical-physical process of generating energy and activity to maintain a constant internal body temperature between 97.7°F and 99.5°F. This physiological mechanism balances the interactive heat exchange between people and their environments.

Metabolic Rate Factors

Age—metabolic rate increases from early childhood to young adult, then drops with age.

Gender—men and women have different *basal* and metabolic activity rates.

Lifestyle—generally active or generally sedentary in lifestyle or work indoors or outdoors.

Table 4.1 Metabolic State—Activity Level

Activity	Heat Produced (BTU/hr)	Watts (W)	100 W Light Bulbs
Sleeping	300	100	[1]
Light work (computer)	600	200	
Walking	900	300	
Jogging	2,400	800	

Table 4.1 compares metabolic activities and the heat they produce. Notice how the increase in heat produced is almost exponential for each nominal increase in activity level.

Metabolic Exchange is how we exchange heat with our surroundings:

Respiration—sensible and latent exchange
Perspiration-latent heat exchange
Convection—sensible heat exchange
Conduction—sensible heat exchange
Radiation—sensible heat exchange

1. IMG 4.1

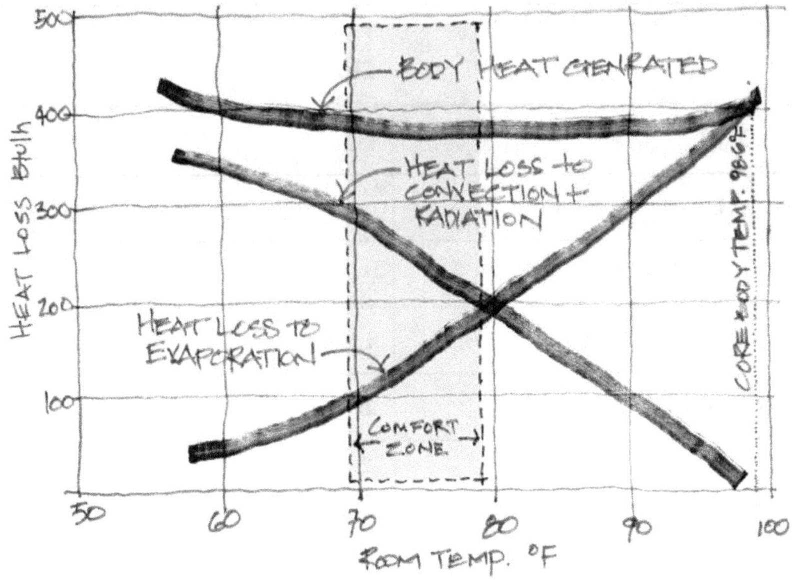

Figure 4.4 Charting the Heat Exchange Between People and Their Thermal Environment Shows How the Mechanism of Exchange Transfers from Convection and Radiation to Evaporation (from Perspiration) as the Ambient Temperature Extends Past the Comfort Zone; the Chart Is an Approximation for a Person at Rest with 45% Relative Humidity.

Clothing

What will people be wearing? Much depends on the activity of course but also the season and the likelihood of a great variation in clothing. ASHRAE (American Society of Heating Refrigeration and Air Conditioning Engineers) has developed a standard rating for clothing to assist in evaluating and calculating the probability of comfort for a variety of dress styles and corresponding environments. The "Clo" is a value assigned to combinations of clothing that might be typical in office environments depending on the season and geographical location. A three-piece business suit, with pants, undershirt, long-sleeve shirt, vest, and jacket has a "clo" rating of 1.00. This clo rating (1.0) is the basis for establishing the ASHRAE comfort zone. A person wearing clothes with a 1.0 clo or similar should be comfortable at the range of temperature and humidity identified as the standard comfort zone.

Figure 4.5 Climate and Seasonal Accommodation Suggest Appropriate Clothing for the Out-of-Doors; This Element Should be Considered When Designing for Comfort in Architectural Spaces.

Clothing	Clo
Shorts, short-sleeve shirt	.36
Pants, short-sleeve shirt	.57
Pants, long-sleeve shirt, jacket	.96
Pants, long-sleeve shirt, jacket, vest	1.00
Knee skirt, short-sleeve shirt, hose	.54
Knee skirt, long-sleeve shirt, full slip, hose	.67
Ankle skirt, long-sleeve, hose, suit jacket	1.10

The "Clo" ratings for various states of dress configurations are shown in the above table as a reference. Minor variations have a limited impact on comfort; however, a clo of 0.5, relative to 1.0, will significantly change the ambient conditions that a person is likely to feel comfortable.

Acclimatization

People become accustomed to their thermal environments over time, in both cases of hot and cold. It has been reported that this can occur over just a couple of weeks. It is true that blood pressure tends to be higher for individuals in colder climates due to constricting of the blood capillaries—a response to cold weather. Contrary to some beliefs, blood viscosity or thickness does not change with climate. People who spend most of their time inside conditioned spaces can become acclimatized to this more controlled environment (indoor acclimatization) and have difficulty being comfortable in outdoor spaces that may be hotter or colder. The reverse, outdoor acclimatization, is also true. People who have lived in moderately hot, humid climates (with no air conditioning) report great difficulty becoming comfortable in air-conditioned spaces due to both the cold and dryness of the air.

Environmental Factors Affecting Thermal Comfort

The natural and built environment affect comfort through air temperature, mean radiant temperature (MRT), air motion, and humidity. Each of these aspects of comfort can be controlled in occupied spaces. Architecture can mitigate the intensity of environmental stresses (hot and cold) by using natural and passive control strategies to provide comfort while limiting the use of energy resources necessary to power the active thermal control systems.

Peter Zumthor's thermal baths in Vols, Switzerland, offers an array of thermal experiences to visitors—very cold, very hot, warm, cool, and neutral. These baths, located in a remote alpine valley, are a major destination for Europeans seeking to relax and recharge suggesting that we are not just seeking static comfort as human beings. Rather, we seek, or perhaps even need, dynamic thermal environments for our well-being. Furthermore, the environments that we spend great amounts of time in will likely become dull and monotonous if they are not somewhat dynamic within the range of general comfort.

Air temperature is a critical factor affecting human comfort. Generally, we are comfortable with air temperatures between around 70°F to 75°F when doing light activities and wearing pants and long sleeves. We lose heat to the air primarily through conduction (being in contact with it), but also through evaporation that is discussed with humidity below. As we get colder, we shiver, restrict capillaries, hunch our bodies, increase body movement, and restrict mental activity. As we get too hot, we dilate our capillaries, perspire, restrict body movement, and restrict mental activity.

Figure 4.6 Peter Zumthor's Design of the Thermal Baths in Vals, Switzerland; Here, Visitors Enjoy Multiple Thermal Conditions from Icy Baths—or Air as in the Case of a Winter Visit—to Steam Rooms with Thermal Change as a Driving Element of the Experience.

The temperature of surfaces and objects around us also have an influence on our comfort through radiant heat exchange. If we are out in the sun on a nice sunny day, a much lower air temperature will be more comfortable than if we are out on a windy night. The radiant energy from the sun warms us. If the walls, floor, and ceiling in a space are warmer than our skin temperature at about 95°F, we will gain heat from those surfaces. Conversely, when room surfaces are cold, we will lose heat to them. Typically, losing heat to space surfaces is more common in winter environments because the outside air can drop to temperatures below freezing. Subsequently, the exterior walls and windows can become quite cold relative to our skin temperature. On the other hand, in hot climates, the interior surface temperatures rarely rise above 95°F. The temperature difference between our skin and nearby surfaces is what drives this radiant heat exchange. If our skin is 95°F and a cold window is 38°F (difference of 57°F). We could lose a lot of body heat to the cold window. However, if the same window is 95°F, then there is no heat exchange between our skin and the window as the temperature difference is zero.

Figure 4.7 Sitting in Front of a Bonfire on a Cold Night, We Can Feel the Radiant Heat from the Fire Warming the Parts of Our Bodies Facing the Fire, While the Other Side Will Remain Cold Both from Radiating Heat to the Cold Night Sky, Cold Surfaces Nearby, and to the Ground; This Heat Loss Is in Addition to Conductive Heat Loss to the Cold Air.

Mean Radiant Temperature (MRT): The arithmetic mean of temperature differences relative to the visual sizes of surfaces surrounding a person.

With regard to warm surfaces as heat sources in winter, people prefer to have the heat arrive from the floor or walls rather than the ceiling. The chart below illustrates high dissatisfaction for warm ceilings and cool or cold walls—which might be indicative of a cold window in a heated room. There is an effective limit on how hot a surface can be and still feel comfortable. A fireplace can be quite delightful when the air is cold but can also overheat the body surfaces facing the fire.

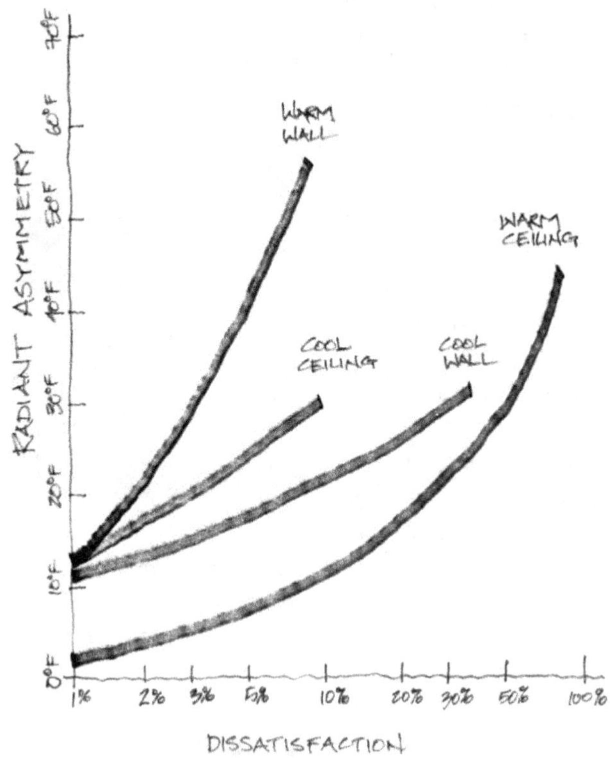

Figure 4.8 The Percentage of People Who Feel Dissatisfied as Radiant Asymmetry Is Increased for Given Architectural Surfaces—Walls and Ceilings; a Warm Ceiling (Relative to Other Surfaces) Approaches 100% Dissatisfaction.

Air movement can significantly affect thermal comfort. The same air temperature and even the same MRT as noted above could be comfortable, hot, or cold, depending on the movement of air. A subtle breeze can transform a hot afternoon into a pleasant one. We may appreciate the value of a ceiling fan or floor fan on a hot afternoon or evening. Conversely, we may have experienced the painful biting of wind gusts on freezing winter days.

Air movement greatly affects our comfort through two heat exchange mechanisms. First, the air blows away the warm layer of air that can build up on the surface of our skin in calm weather, thus dramatically increasing the conductive heat loss to the air. Second, it increases the effectiveness of evaporation by blowing away any moisture saturated air from our perspiration and providing new, dryer air to cool us further. As moisture evaporates from our skin, it provides a very effective cooling mechanism. Air movement can be an effective cooling strategy, but if the ambient temperature is too low, moving air can feel drafty and chilling even at low levels of air movement.

In warm to hot environments air movement is typically achieved with floor or ceiling fans and can provide comfort in temperatures up to 86°F at 50% relative humidity. This is the equivalent of reducing the ambient temperature by 6°. In the charts below, 83°F is plotted against still air (left) and air moving at 100 feet per minute (fpm). The blue band shows the area of comfort based on subjective research studies with numerous individuals. For warmer temperatures above 70°, air movement between 50 and 200 fpm is generally considered pleasant, but noticeable beginning at rates of 100 fpm. Common ceiling fans can provide air speeds of up to 450 fpm when set on high. Clearly, with moderate air movement we can be comfortable at higher temperatures and reduce the mechanical cooling required in air-conditioned buildings. As of 2022, ASHRAE has accepted the CBE Comfort Tool (discussed below) to model comfort and air movement and is working on improving the standards for air movement as part of their ASHRAE 55 building standard.

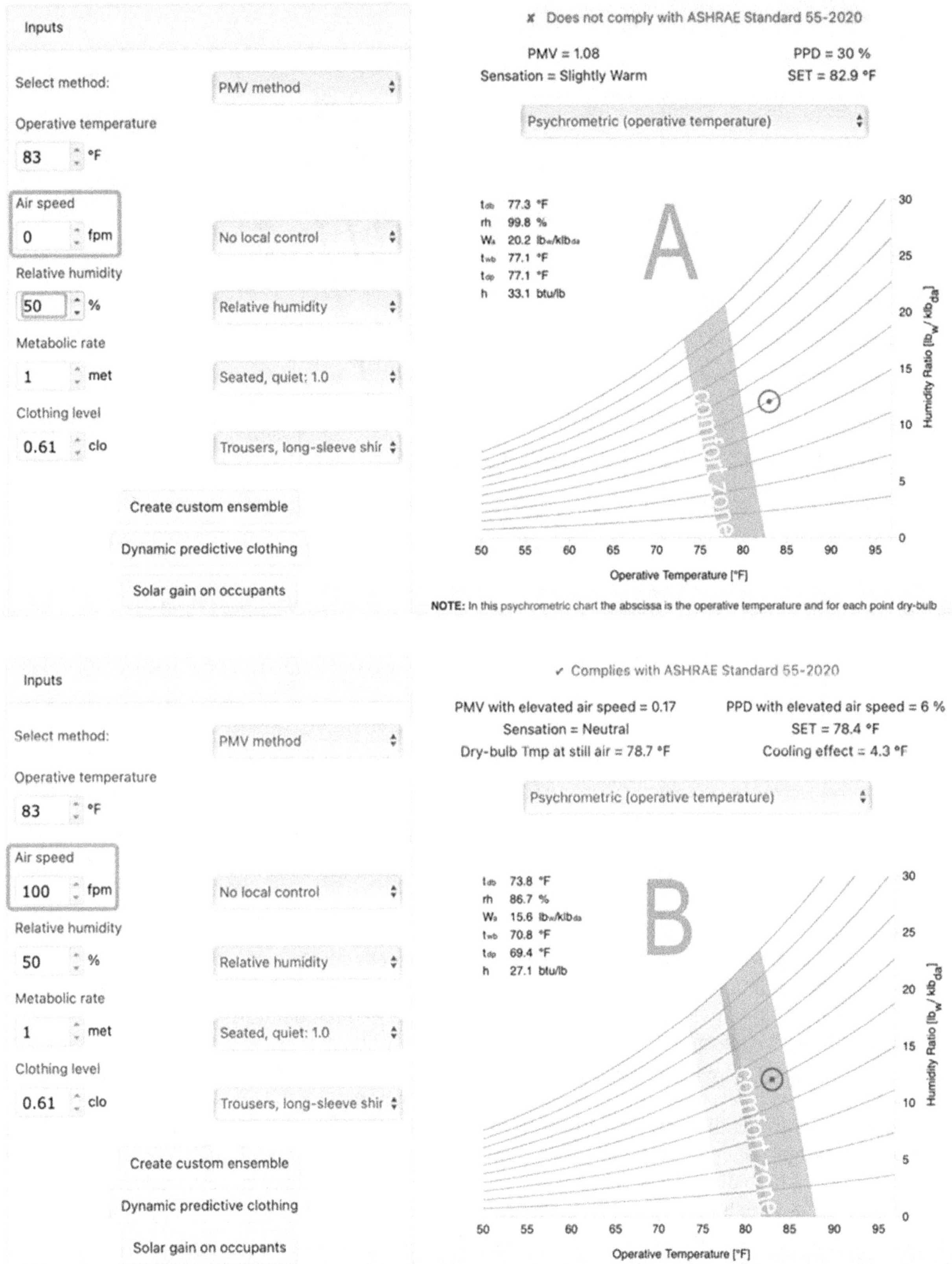

Figure 4.9 Center for the Built Environment (CBE Comfort Tool) https://comfort.cbe.berkeley.edu; the Upper Thermal Condition (A) Shows a Point Outside the Comfort Zone (83°F, 50%rh, 1 met, 0.61 clo) for a Person Doing Sedentary Activity Who Is Casually Dressed; Adding Air Movement of 100 fpm (Low Ceiling Fan Setting) Effectively Moves the Comfort Zone into a Warmer Range of Temperatures as Shown in Chart B.

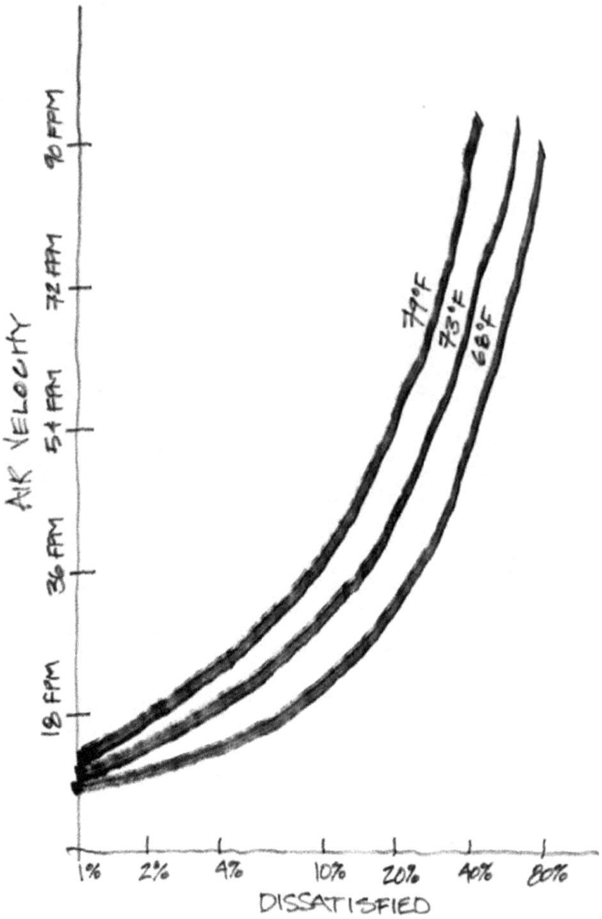

Figure 4.10 Chart of Satisfaction/Dissatisfaction for Air Movement for Common Temperatures; Reading This Chart, One Can See That Dissatisfaction Increases with Air Movement as the Temperature Decreases; When the Air Is Cooler or Cold, We Perceive It as Chilly or Drafty; However, at Higher Temperatures, Air Movement Is Typically Welcomed.

Take a few minutes and go to the CBE Comfort Tool site and test some scenarios of your own for their comfort prediction. Does the prediction correspond to your experiences?

From the National Oceanographic and Atmospheric Administration (NOAA) chart above, wind chill or moderate breezes of 5–10 miles per hour can reduce the perceived temperature by 4°F to 6°F, and as the temperature drops, the effect becomes much more extreme. Although these representative temperatures are much lower than what we might find in buildings, it reinforces the concept that air movement should be limited to very low levels to achieve comfort in cool environments.

Humidity describes the amount of moisture in the air and more precisely, the *relative humidity.* Because the physical volume of air changes with temperature while the amount of water vapor in the air stays the same, the humidity is relative to the temperature. Relative humidity is measured as a percentage of water vapor relative to how much water the air can hold—60% relative humidity, for example. This is discussed in more detail in the module on psychrometrics.

Wind (mph) \ Temperature (°F)	40	35	30	25	20	15	10	5	0	-5	-10	-15	-20	-25	-30	-35	-40	-45
Calm	40	35	30	25	20	15	10	5	0	-5	-10	-15	-20	-25	-30	-35	-40	-45
5	36	31	25	19	13	7	1	-5	-11	-16	-22	-28	-34	-40	-46	-52	-57	-63
10	34	27	21	15	9	3	-4	-10	-16	-22	-28	-35	-41	-47	-53	-59	-66	-72
15	32	25	19	13	6	0	-7	-13	-19	-26	-32	-39	-45	-51	-58	-64	-71	-77
20	30	24	17	11	4	-2	-9	-15	-22	-29	-35	-42	-48	-55	-61	-68	-74	-81
25	29	23	16	9	3	-4	-11	-17	-24	-31	-37	-44	-51	-58	-64	-71	-78	-84
30	28	22	15	8	1	-5	-12	-19	-26	-33	-39	-46	-53	-60	-67	-73	-80	-87
35	28	21	14	7	0	-7	-14	-21	-27	-34	-41	-48	-55	-62	-69	-76	-82	-89
40	27	20	13	6	-1	-8	-15	-22	-29	-36	-43	-50	-57	-64	-71	-78	-84	-91
45	26	19	12	5	-2	-9	-16	-23	-30	-37	-44	-51	-58	-65	-72	-79	-86	-93
50	26	19	12	4	-3	-10	-17	-24	-31	-38	-45	-52	-60	-67	-74	-81	-88	-95
55	25	18	11	4	-3	-11	-18	-25	-32	-39	-46	-54	-61	-68	-75	-82	-89	-97
60	25	17	10	3	-4	-11	-19	-26	-33	-40	-48	-55	-62	-69	-76	-84	-91	-98

Frostbite Times: 30 minutes; 10 minutes; 5 minutes

Wind Chill (°F) = 35.74 + 0.6215T - 35.75($V^{0.16}$) + 0.4275T($V^{0.16}$)

Where, T= Air Temperature (°F) V= Wind Speed (mph) *Effective 11/01/01*

Figure 4.11 National Oceanographic and Atmospheric Administration (NOAA) Wind Chill (Air Movement) Effect on Perceived Temperature.

As we try to give up heat to the air, to maintain comfort, we may start to perspire, and we may not notice this unless the air is humid and unable to accept the moisture we are trying to add to the air. When we begin to perspire, moisture on our skin evaporates giving up heat at a rate about five times greater than from conductive contact with the air. If the air is dry and warm, we may not even notice that we are perspiring. However, if the air is humid, say 75%–100% relative humidity, the ability to evaporate the moisture on our skin will be greatly reduced and we will become hot and sweaty. Perspiration is a critical physiological response to avoid overheating when we unable to give up enough heat to the air through contact (conduction). When becoming warm, the body initiates eccrine sweat glands—the most numerous, productive, and located over most of the body. Interestingly, these may also be initiated when we eat spicy food or digest a heavy meal. In the case of spicy food, our body thinks it is overheating as the spice activates our sense of heat. In the case of digestion, this requires extra metabolic activity, generating heat and the need to cool the body. Perhaps restaurants, especially those serving spicey foods should have lower ambient temperatures than other spaces for nominally sedentary activity.

NWS Heat Index

Relative Humidity (%) \ Temperature (°F)	80	82	84	86	88	90	92	94	96	98	100	102	104	106	108	110
40	80	81	83	85	88	91	94	97	101	[illegible]	109	114	119	124	130	136
45	80	82	84	87	89	93	96	[illegible]	104	109	114	119	124	130	137	
50	81	83	85	88	91	95	99	103	108	113	118	124	131	137		
55	81	84	86	89	93	97	101	[illegible]	112	117	124	130	137			
60	82	84	88	91	95	100	105	[illegible]	116	[illegible]	129	137				
65	82	85	89	93	98	103	108	114	[illegible]	[illegible]	136					
70	83	86	90	95	100	105	112	119	126	[illegible]						
75	84	88	[illegible]	97	103	109	116	124	132							
80	84	[illegible]	94	[illegible]	106	113	121	129								
85	85	90	96	102	110	117	126	135								
90	86	91	98	105	113	122	131									
95	86	93	100	[illegible]	117	127										
100	87	[illegible]	103	[illegible]	[illegible]	[illegible]										

AM PM

Likelihood of Heat Disorders with Prolonged Exposure or Strenuous Activity

Caution; Extreme Caution; Danger; Extreme Danger

Figure 4.12 National Oceanographic and Atmospheric Administration (NOAA) Wind Chill (Air Movement) Effect on Perceived Temperature.

The figure above identifies with white circles typical morning and afternoon conditions in hot humid climates such as Louisiana or Florida. From the chart, one can follow the changes as the air heats up during the day, the relative humidity decreases (p.m. condition), while in the morning (a.m. condition), the air is somewhat cooler with a relative humidity that is much higher. This is a result of the air volume changing with temperature even though the same amount of moisture is in the air. In both cases, the afternoon and morning ambient temperatures would be quite uncomfortable, even with breezes, because of the humidity. Even when the air is cooler around 68°F to 75°F, too much humidity can be perceived as being clammy or uncomfortable.

When heating in cold seasons or in arid desert environments, the air can also become too dry with relative humidity dropping below 30%. This too causes discomfort as we lose too much moisture to evaporation causing dry skin, eyes, and noses.

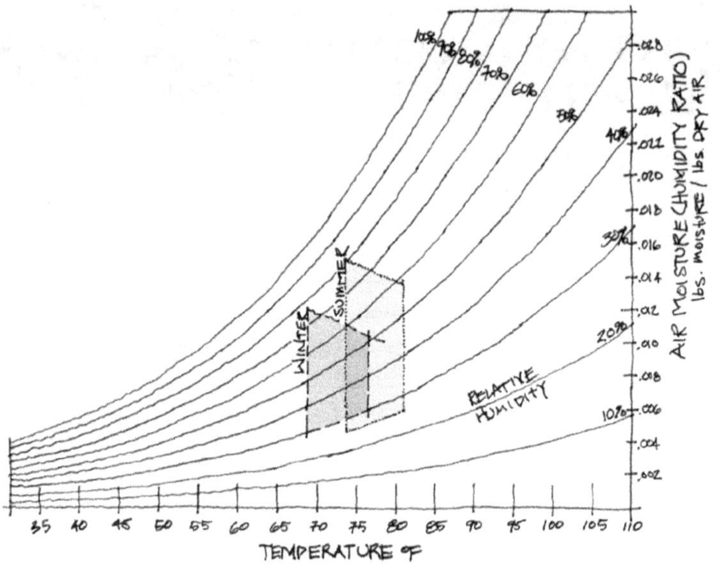

Figure 4.13 Simplified Psychrometric Chart Illustrating the Relationship Between Air Temperature, Moisture, Relative Humidity, and Comfort Zones for Both Winter And Summer Clothing and Acclimatization; the Comfort Zones Assume Sedentary Activities such as Sitting and Working at a Desk.

The Comfort Zone

Using the concepts introduced above, researchers from ASHRAE developed the concept of a comfort zone—a range of temperatures and humidity that people will find comfort in addition to adjusting for summer and winter acclimatization. The comfort zone is an area plotted on the simplified diagram of ASHRAE's psychrometric chart below showing comfort and areas that are too hot or too cold. The psychrometric chart maps the typical conditions of moist air with the corresponding ambient temperature (°F) and moisture in the air (%rh).

Originally developed in the 1970s, the comfort zone represents a range of thermal conditions in which most men working in office attire would be comfortable. Continued research and more advanced models for charting the comfort zone have been adopted in 2020 as an addendum to the ASHRAE-55 standard. The updates seek to include low energy strategies such as air movement, refine the predictability of human perceptions of comfort, and to model more casual clothing that is now typically worn in office environments. It is important to note that the comfort zone—although offering a range of conditions—is still based on a fundamentally limited set of circumstances and must be interpreted to be appropriately used as a reference for situations that differ from the original model.

Comfort Zone Factors

Person: adult male

Activity: office activity (sedentary)

Apparel: wearing a business suit (clo of 1.0)

MRT: surfaces are the same temperature as the ambient air temp

Air movement: limited to less than 50 fpm.

The comfort zone provides a great reference for understanding and predicting comfort and evaluating the potential value of passive design strategies toward achieving comfort and reducing the need to mechanically control the thermal environment. The internet tool *Climate Consultant* plots annual hourly temperatures on the psychrometric chart for any weather station located on the globe against the ASHRAE comfort zones (winter and summer clothing), providing a basis for evaluating the climate and measuring the effectiveness of passive and active thermal control systems. The tool charts the number of hours of comfort that might be provided by passive and active heating or cooling strategies. The tool is quite useful for students and practitioners to prioritize and evaluate design strategies to provide thermal comfort within their respective climates.

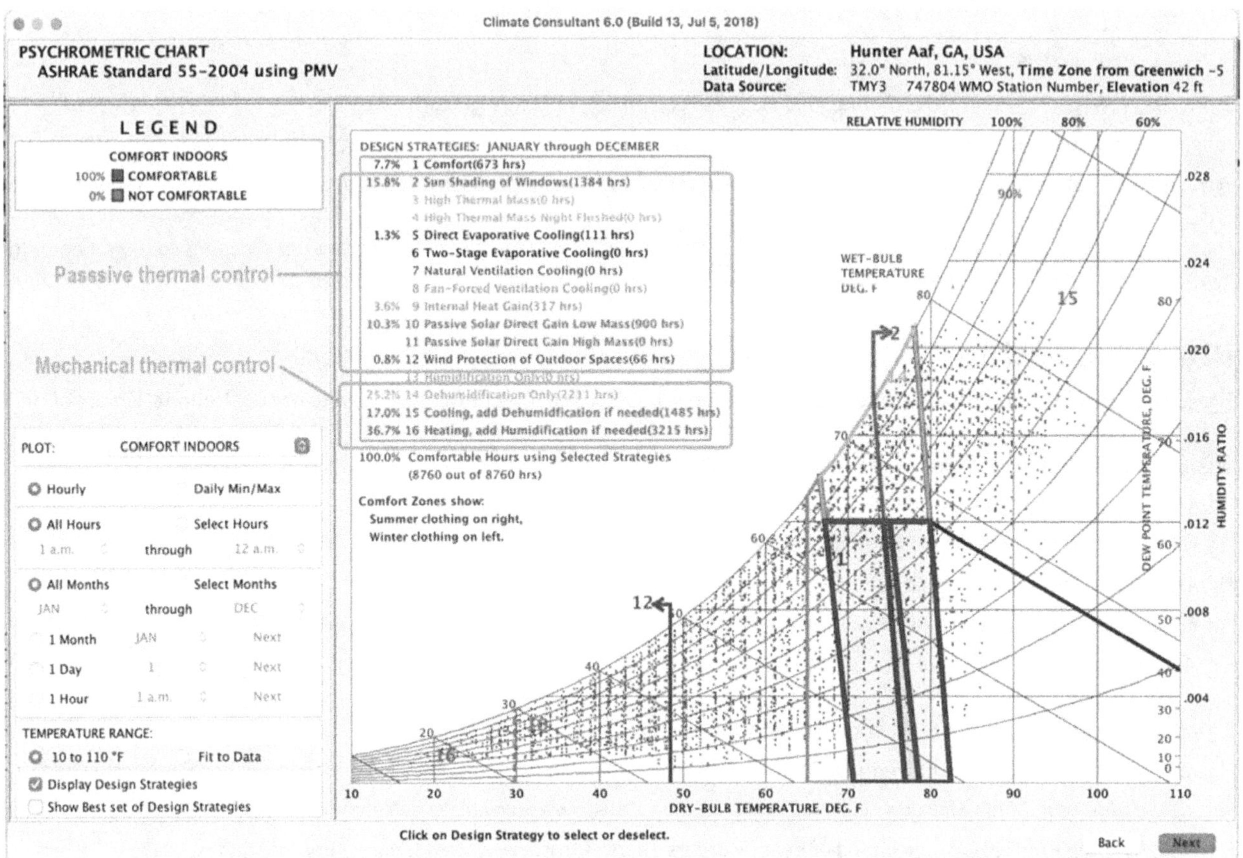

Figure 4.14 Psychrometric Chart with Hourly Temperature and Humidity Data Plotted (Dots) Against the Summer and Winter Comfort Zone; Users Can Select Which Passive and Active Mechanical Strategies They Would Like to Use and Measure the Percentage of the Year Comfort Is Provided; the Configuration Shown Provides 100% Comfort Using Both Passive and Active Mechanical Systems.

The software Climate Consultant can be downloaded at:

www.sbse.org/resources/climate-consultant

Summary

Our relationship with the thermal environment, comfort or discomfort, is fundamental to our well-being, enjoyment, and to our productivity when working. When we use energy to mechanically heat or cool spaces in our architecture, we want to be sure that it is in the service of comfort, and that we are not over cooling or over heating people. Additionally, we should be keen to utilize passive strategies whenever possible to extend the time that comfort is naturally provided within the climate we are occupying. For most climates, there are multiple times of the year that are naturally comfortable. It makes sense that would not expend energy to thermally control spaces during those times. Exploring how we might achieve that through our architecture provides a great opportunity for clever designers.

References

American Society of Heating Refrigeration and Air-Conditioning Engineers, Inc. (2001). *ASHRAE handbook, fundamentals*. American Society of Heating Refrigeration and Air-Conditioning Engineers, Inc.

Arens, E. (PI), Tartarini, F., Schiavon, S., Cheung, T., & Hoyt, T. (2020). CBE thermal comfort tool: Online tool for thermal comfort calculations and visualizations. Center for the Built Environment. https://comfort.cbe.berkeley.edu

ASHRAE. (2020). Thermal Environmental Conditions for Human Occupancy. ANSI/ASHRAE Addendum d to ANSI/ASHRAE Standard 55-2017. https://www.ashrae.org/file library/technical resources/standards and guidelines/standards addenda/ 55_2017_d_20200731.pdf

Baker L. B. (2019). Physiology of sweat gland function: The roles of sweating and sweat composition in human health. *Temperature*, *6*(3), 211–259.

Climate Consultant, US Department of Energy.

Grondzik, W., & Kwok, A. (2019). *Mechanical and electrical equipment for buildings* (13th ed.). Wiley.

Lang, S. (2004, October). *Study links warm offices to fewer typing errors and higher productivity*. Cornell Chronicle.

Pryor, J. A., & Prasad, A. S. (2008). *Physiotherapy for respiratory and cardiac problems: Adults and paediatrics* (p. 8). Elsevier Health Sciences.

Zhang, H., Pasut, W., Zhai, Y., Hoyt, T., & Huang, L. *Air movement as an energy efficient means toward occupant comfort*. Center for the Built Environment, Wurster Hall University of California.

Image Credits

Fig. 4.2: Adapted from Food and Nutrition Board, "Metabolic Rate for Activities by Gender." Copyright © by National Library of Medicine.

IMG 4.1: Copyright © by Microsoft. Reprinted with permission.

Fig. 4.5a: Copyright © 2018 Pexels/Orest Sv.

Fig. 4.5b: Copyright © 2019 Pexels/ganimatque.

Fig. 4.5c: Copyright © 2024 Depositphotos/HayDmitriy.

Fig. 4.7: Copyright © 2017 Pixabay/StockSnap.

Fig. 4.8: The American Society of Heating, Refrigerating and Air-Conditioning Engineers (ASHRAE), "Radiant Asymmetry Graph." Copyright © 2017 by Techstreet.

Fig 4.9a: Copyright © 2023 by CBE Thermal Comfort Tool. Reprinted with permission.

Fig 4.9b: Copyright © 2023 by CBE Thermal Comfort Tool. Reprinted with permission.

Fig. 4.11: NOAA, "Wind Chill Effect on Perceived Temperature," https://www.weather.gov/safety/cold-wind-chill-chart, 2001.

Fig. 4.12: Adapted from NOAA, "Likelihood of Heat Disorders Chart," https://www.weather.gov/ama/heatindex, 2017.

Fig. 4.14: Regents of the University of California, "Psychrometric Chart Temperature and Humidity." Copyright © 2023 by The Society of Building Science Educators.

Module 5

Thermal Properties of Air & Psychrometrics

Martin Gold

The air is so thick you could cut it with a knife!

Introduction

The idiom above has been attributed to many situations—one of the more common is being confronted with muggy or foggy air. Maybe you've heard the phrase "*but it's a dry heat,*" suggesting that high temperatures are more tolerable with low humidity. One might conclude from these cultural idioms that people have an intuitive understanding that humidity (moisture in the air) affects our comfort independently of the air temperature.

There are five mechanisms of heat flow that affect human comfort, and the amount of energy needed to heat or cool architectural spaces. The relationship between air temperature and the amount of moisture in the air (humidity) is somewhat complicated and can be critically important in terms of thermal comfort and maintaining healthful air quality. For mechanical engineers, the study of moist air is referred to as ***psychrometry***. Air almost always comprises a combination of water vapor and dry air and the ratio of air and water vapor is quite dynamic and varies as air temperature is increased or reduced. Considering thermal comfort, we exchange heat with air around us, and our ability to do that depends on how much heat there is in the air at any given time. This heat can be present in the air temperature (sensible heat) or in the amount of moisture in the air, referred to as latent heat.

Figure 5.1 Vernazza, Cinque Terra, Italy on the Ligurian Sea; Here We Have a Dramatic Image Suggesting the Energy Exchanges of Solar Radiation, Evaporation, and the Potential of Laten Heat Contained in the Moisture in Air.

Thermal Environment

In this module, we will be discussing the ecology of heat in the air as it is sensed (sensible heat) and the latent heat—the heat from the moisture that is in the air. The image below is captured in Vernazza, Italy, of the five cities referred to as Cinque Terra, overlooking the Ligurian Sea, the image captures the dramatic interrelationship of sun, air, water, and land as interrelated elements of heat exchange as a thermal ecology. The Ligurian Sea, warmed by the sun, evaporates water (water to water vapor) adding humidity to the air. The sun's radiant energy also warms the land.

As the humid sea air meets the land, conductive heat exchange warms the moist air. This creates convective flow causing the warm moist air to rise. At higher altitudes, the air cools causing the water to condense onto particles in the air forming clouds. When enough moisture condenses, the mass of water droplets will become heavy enough for it to fall as rain.

Five Factors of Heat Flow

1. **Sensible heat (air temperature)**
2. **Latent heat (moisture in the air)**
3. Convection (air movement)
4. Conduction (objects in contact)
5. Radiation (objects in view)

Factors in bold are covered in this module and factors 3–5 will be covered in the next module on thermal properties and heat flow.

Heat Flow

When we are trying to understand these principles of heat flow and heat exchange between and within materials, it is critically important to keep in mind that ***heat moves naturally from high concentration to low concentration***. Heat moves from hot to cold. When we are discussing heat flow, we are always following the heat. For example, when we are cooling spaces, we should conceptually be thinking about removing the heat rather than cooling the air. This cognitive framework or "removing heat" is an accurate model of how the heat flows. If we are "removing the heat", we can then ask the question: where are we putting the heat? The heat we remove must go somewhere.

Latent Heat

Water vapor in the air—latent heat—is holding heat energy from the transformation of liquid water to water vapor. For a material to change state, liquid to vapor, or solid to liquid, heat is absorbed by the medium. Often, the heat comes from the surrounding air or perhaps from sunlight. When a material changes state from vapor to liquid, or from liquid to solid (ice), heat is given up or emits heat to the surrounding environment. Liquids, solids, and vapors are considered "phases" of a material and when a material changes phase, such as when ice freezes (from liquid to solid) it gives up a quantity of heat in the process. When water evaporates, the process must absorb heat to change the phase from liquid to vapor. The absorbed heat, required to change the phase of the material, remains present in the water vapor—suspended in the air as "latent heat". The latent heat will be released when the water condenses back into a liquid.

For example, if we start to boil a simple pot of water (1 pound of water to be exact), it takes about 180 British Thermal Units (BTU) to raise the temperature of pot of water to the boiling point of 212°F. At this temperature, the water will no longer get hotter. If you keep a thermometer in the water and keep heating it, the temperature will not rise above 212°F. This phenomenon also makes cooking times reliable, as boiling water is reliably steady at 212°F. The temperature does vary slightly with atmospheric pressure, so things are a bit different at different high altitudes. If we keep heating the water in the pot, it starts to boil, and then the water will not get hotter, but it will evaporate as it boils. When all the water has evaporated, it will have absorbed 1,061 BTUs of heat. This is the heat energy required to change the water from a liquid to a gas in the evaporation process. And, that latent heat (1,061 BTUs), is now in present in the air as moisture or added humidity.

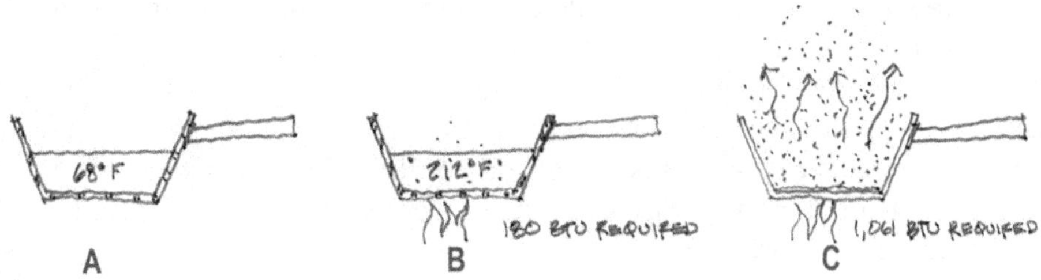

Figure 5.2 Diagramming the Boiling a Pot of Water to Illustrates the Phenomenon of Phase Change from Water to Vapor, and the Latent Heat That Is Embodied in the Water Vapor Suspended in Air;

When we cool spaces mechanically, both sensible heat (air temperature is reduced) and the latent heat (heat in the moisture in the air) is removed from the air. Perhaps you have noticed that air conditioning systems often emit water, referred to as condensate, which is the result of the moisture condensing out of the air when the heat is removed (cooled) by the system. In very humid climates, this latent heat can use up to six times more energy to remove than just lowering the sensible air temperature. Laten heat is a substantial part of the heat in air especially when humidity is high.

Sensible Heat

This is quite simply the heat that we sense or feel through our skin—what we feel when the air in contact with our bodies is cool or warm. Skin temperature is approximately 97°F so if the air is cooler than skin temperature, it generally feels moderate to cool, and we are losing heat to it. If the air is warmer, we are gaining heat. Just to be very clear, if the air temperature is the same as our skin temperature, no heat exchange would occur. When we measure the air temperature with a typical mercury thermometer, or if we check our smart phones, or the internet for the temperature, we are getting a measure of the sensible heat.

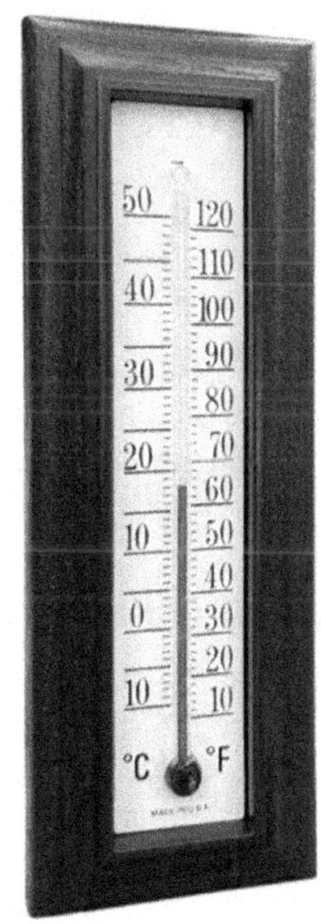

Figure 5.3 The Sensible Temperature That We Perceive Through Contact with the Air (Sensed) Is Measured with a Mercury Thermometer.

Air Volume, Temperature, and the Capacity to Hold Moisture

All materials expand when they are heated and contract when they are cooled including the air around us. Heat energizes the molecules in air and the molecules push harder on each other and the surroundings effectively expanding. If the air is held in a container, then the pressure inside that container will increases as the air is heated because the air is contained and being prevented from expanding.

There is an interesting interactive demonstration of the interaction of heat and molecular activity at the following link: States of Matter (PhET Interactive Simulations)

When you get to the link follow these steps:

- Click on Phase Changes
- Select "Water" under Atoms and Molecules
- Click on the green sign to open the phase diagram
- Add/remove heat, pressure (bicycle pump), or reduce the space (press the finger down)

See how the molecules of water change state in responce to pressure and heat.

When air is heated, it expands increasing the air volume—the molecules get farther apart. As the air volume increases, it can also hold more moisture in the form of water vapor. The inverse is also true. When air is cooled, the volume is reduced, and it can hold less water vapor. When we cool humid air, at some point, the air can no longer hold all the water vapor, and the moisture in the air condenses back to water, typically on cool surfaces. It is why a cold bottle of soda or drink glass gets wet on the outside. This concept of water condensing out of the air is discussed in more detail later in this module but it is an important issue for health and durability in architecture.

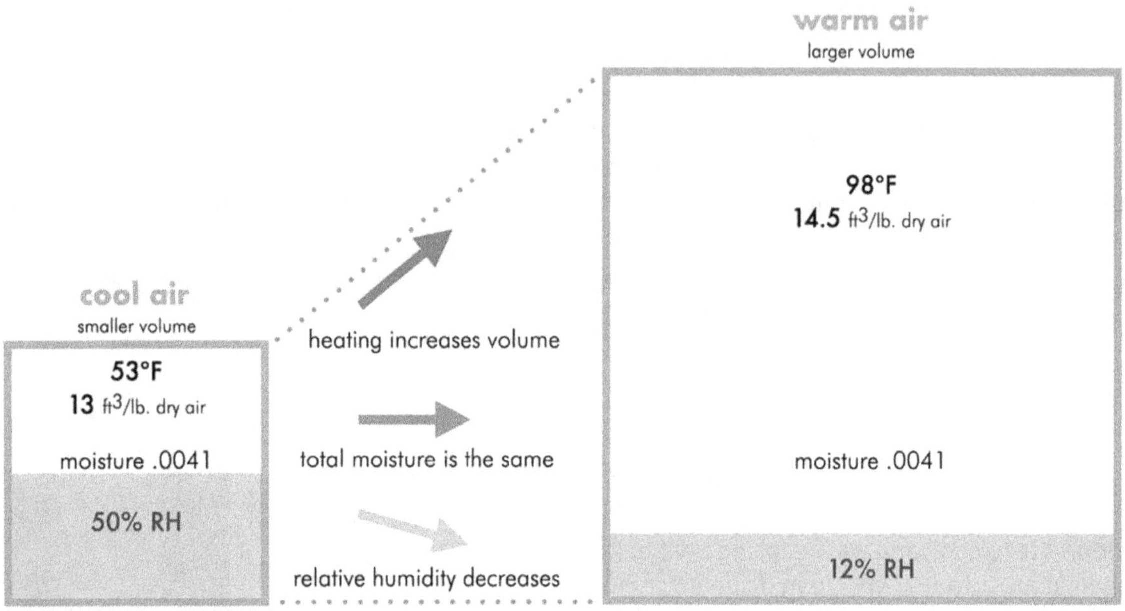

Figure 5.4 Diagram Illustrating the Relationship of Air Volume, Temperature, and Relative Humidity (rh); the Red Box Shows a Volume of Air at 53°F, 13 Cubic Feet per Pound of Dry Air (the Volume) and 50% rh Being Warmed to a Temperature of 98°F; In This Case, the Volume Increases to 14.5 Cubic Feet per Pound of Dry Air and the Relative Humidity Drops to Only 10%; We Will Track This Process as We Introduce the Psychrometric Chart Later in This Module.

Relative Humidity

Looking at the diagram above in Figure 5.4, we can reference the smaller box, representing an air sample at 53°F and 50% relative humidity (rh). In this example, the air is holding 50% of its total capacity (100%) of holding water vapor. If we 'sensibly' heat the air (only change the air temperature) to 98°F, we can see the box representing the air is larger, illustrating the volumetric expansion of the air due to heating. The 53°F box of air has a volume of 13 cubic feet per pound of dry air. When we heat the air to 98°F, the volume of the air increases to 14.5 cubic feet per pound of dry air. Again, when we heat air, it expands. We can track these relationships on the Psychrometric Chart that will be introduced later in this module.

Although the temperature and subsequently the volume of the air has changed, the total amount of water vapor in the air sample has remained the same—no moisture was added or removed. So, at 98°F, the water vapor in the air is now a smaller proportion of the total air volume than it was when the air was at 53°F. The amount of water vapor did not change, but the size of the container we might say (the air) got larger. This effectively reduces the percentage of water relative to the volumetric size of air. Moisture in the air is referred to as the 'relative humidity' because it is relative to the temperature.

We might not notice the humidity when it is 50%, typically that is quite comfortable. But if when we heat air that is 53°F at 50% rh to 98°F at 10% rh, we would think it is too dry, and it may make our skin, eyes, or nasal cavity dry and itchy. In this scenario, the total amount of water in the air was unchanged, however the perceived comfort—relative to the moisture in the air—is significantly different changing from comfort to too dry.

It is quite common in cold climates to have very dry air in households because the outside air is generally cold and somewhat dry. When we heat that cold dry air, typically to 130°F, it dramatically reduces the relative humidity resulting in air that is very dry. To improve comfort, many people use humidifiers in the winter to add moisture back to the air.

Psychrometric Chart

This chart provides a valuable tool to track the relationship between moisture and air. Prior to computer algorithms, the Psychrometric Chart was the primary tool to track the effects of air conditioning and to design mechanical cooling systems. The American Society of Heating Refrigeration and Air Conditioning Engineers (ASHRAE) is the organization that sets standards for heating, cooling, and ventilating buildings and typically their standards are adopted by municipalities and government agencies when setting codes for the thermal control of buildings. The ASHRAE Psychrometric Chart, adapted below, maps the relationships between sensible heat, moisture in the air, relative humidity, enthalpy, and the saturation point (also called dew-point temperature) for a given air sample. Each of these measures of the heat content in air are important for architects who seek to be ecologically sensitive in their design, reduce reliance on external energy, and who seek to provide a range of comfortable environments for those inhabiting their architecture.

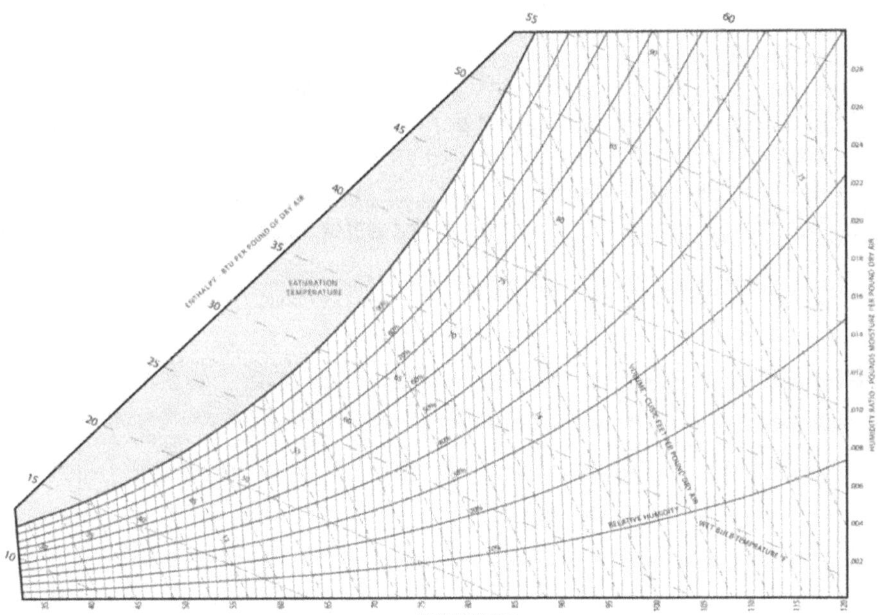

Figure 5.5 Simplified Primary Elements of the ASHRAE Psychrometric Chart: Dry Bulb Temperature; Wet Bulb Temperature; Relative Humidity; Moisture; Enthalpy; Dew-Point Temperature.

Sensible heat is the heat we feel on our skin as we move through the air in a space or outside. When we measure sensible heat, we typically use a mercury thermometer as illustrated in Figure 5.3. The ASHRAE Psychrometric Chart refers to this temperature as the "dry bulb temperature" measured in degrees Fahrenheit (°F). When measuring the air for both temperature and relative humidity, two thermometers must be used: (1) a dry bulb thermometer (a typical thermometer) and (2) a wet bulb thermometer that has moistened gauze or fabric surrounding the bulb of a typical thermometer. The difference between those two temperatures (wet bulb and dry bulb) is used to determine the relative humidity.Figure 5.6 highlights the dry bulb temperature along the "x" axis at the bottom of the Psychrometric Chart. The range of temperatures shown is from 35°F to 120°F.

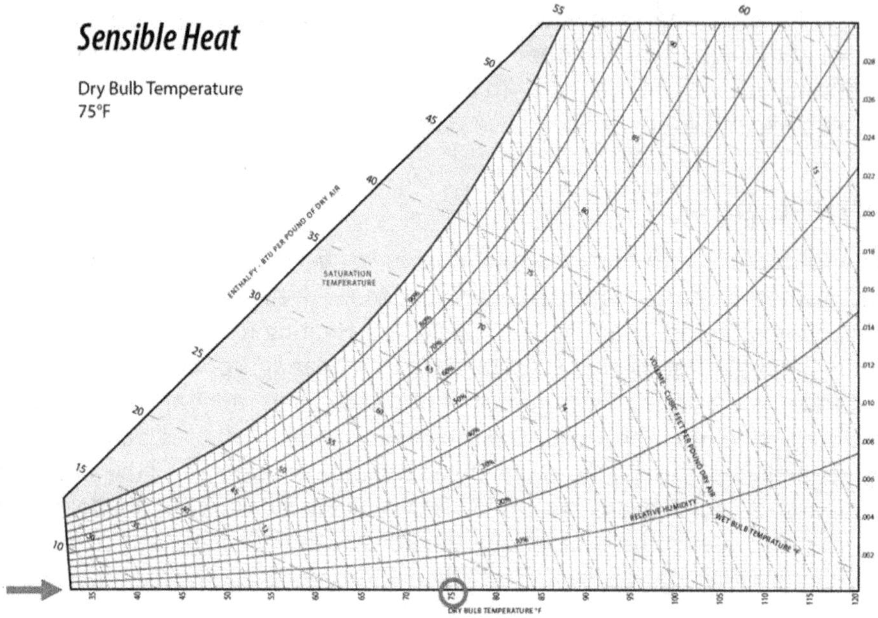

Figure 5.6 Sensible Heat Is Measured Along the Bottom Axis of the Psychrometric Chart Titled Dry Bulb Temperature with Temperatures Ranging from 35°F to 120°F.

Latent Heat is present in the air in the form of water vapor that is moisture in the air. This is measured as the humidity ratio in units of pounds of moisture per pound of dry air. In the air sample example that is charted in Figure 5.7, a typical indoor air temperature of 75°F and 54% relative humidity would have 0.010 pounds of water per pound of dry air. When we perspire, or boil water, or even mist water into the air, the water can evaporate into the air increasing the amount of water vapor and that can be measured by the humidity ratio.

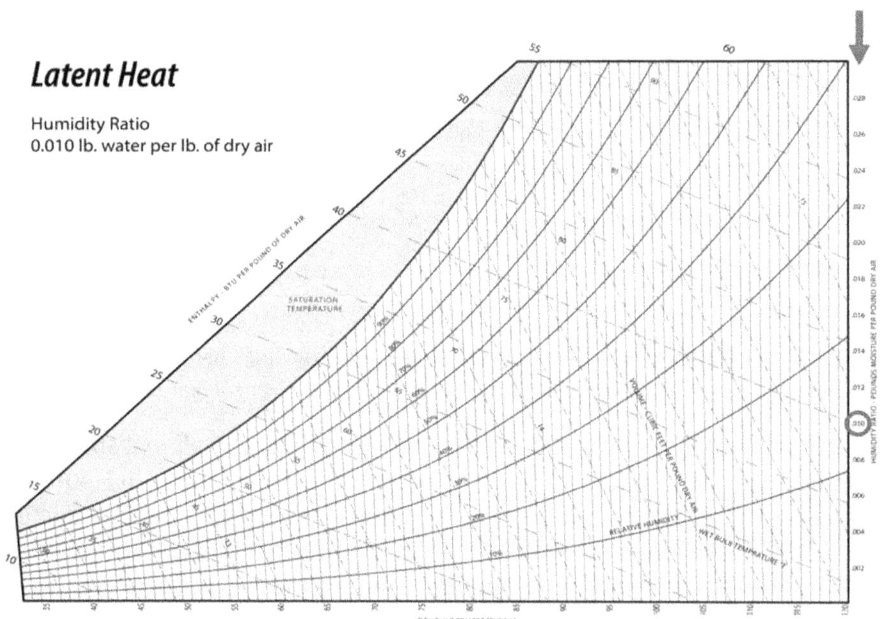

Figure 5.7 Latent Heat Is Measured as the Humidity Ratio in Units of Pounds of Moisture per Pound of Dry Air on the Psychrometric Chart; as Moisture Is Evaporated into the Air, the Humidity Ratio Will Increase and as Moisture Is Condensed Out of the Air, the Humidity Ratio Will Decrease.

Relative Humidity measures the proportion of moisture in the air relative to how much moisture the air can hold at a given temperature. For example, if an air sample is at 100% relative humidity (rh), it cannot absorb any more moisture, and if we are experiencing that condition, and begin to perspire, the air will not allow the moisture on our skin to evaporate, and we will become sweaty. If the air sample has perhaps 20% rh, and we feel hot and begin to perspire, likely we will not even feel the moisture on our skin as it will readily evaporate into the dry air around us. Relative humidity is represented as curved lines on the Psychrometric Chart as highlighted in Figure 5.8.

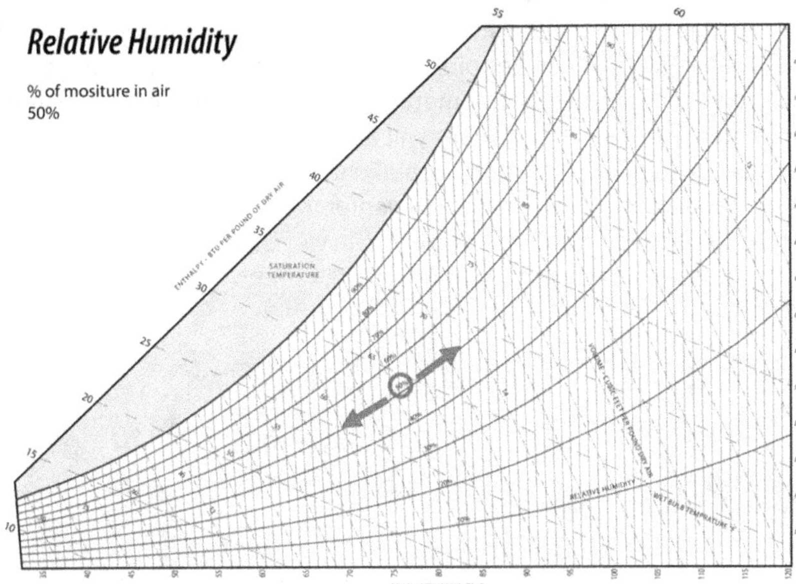

Figure 5.8 Relative Humidity (rh) Is Tracked Using the Curved Lines Arcing Upward from Left to Right on the Psychrometric Chart Ranging from 10% rh to 90% rh; When We Approach 100% rh, We Reach the End of the Psychrometric Chart Representing the Point of Saturation or the Dew Point—Causing Moisture to Condense on Surfaces.

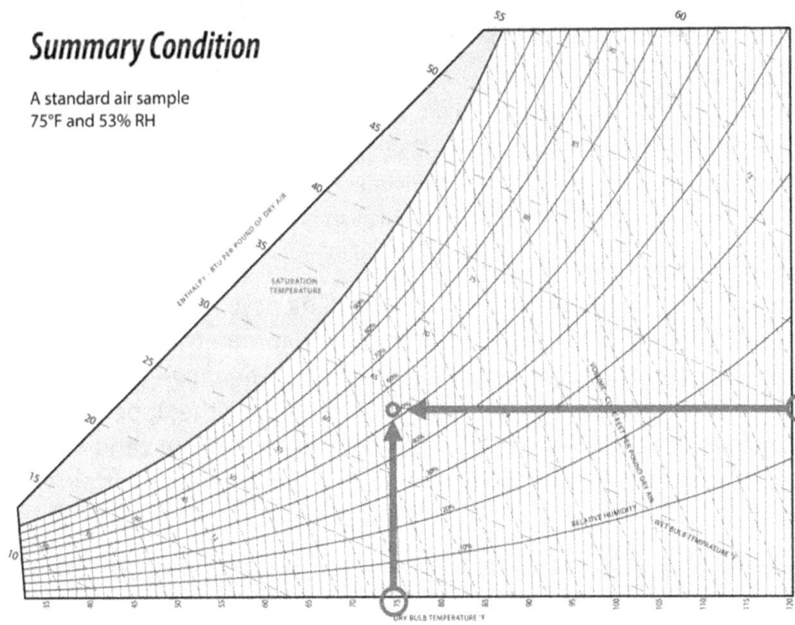

Figure 5.9 The Humidity Ratio and Dry Bulb Temperature Are Plotted on the Psychrometric Chart to Determine the Relative Humidity (rh) for a Given Sample of Air; in This Case, the Air Has 53% rh.

If we have an air sample with the following conditions: Humidity Ratio (0.010 lb. water/lb. dry air) and the dry bulb temperature (75°F), we can plot those measurements to the point at which they intersect on the Psychrometric Chart and find the relative humidity (rh). In the case, we find that the air sample would have a relative humidity of 53%. What would the relative humidity be if we add 0.005 lb. water/lb. dry air to the same air sample? See if you can chart that. The answer is at the end of the module.

Practically, we usually determine the humidity ratio and the relative humidity by physically measuring the air with a **sling psychrometer**. This device has within it, a dry bulb thermometer and the wet bulb thermometer. The wet bulb thermometer, as is implied, is wet. The bulb of the **wet bulb thermometer** is wrapped in gauze or fabric that is moistened with water just prior to taking a measurement. The gauze is just wet enough that it is saturated but not dripping. The device, with two thermometers (wet and dry), allows the measurer to "sling" or spin the thermometers through the air taking a reading of the temperature with each thermometer. If the air is dry, there will be significant evaporation at the wet bulb thermometer. As you might recall, when liquid evaporates and changes state to a vapor absorbing heat energy. Subsequently, the wet bulb thermometer will have a lower temperature than the dry bulb thermometer. If the air is quite moist, much less evaporation will occur at the wet bulb thermometer and the two temperatures (wet bulb and dry bulb) will be nearly the same. At 100% relative humidity, both thermometers will register the same temperature.

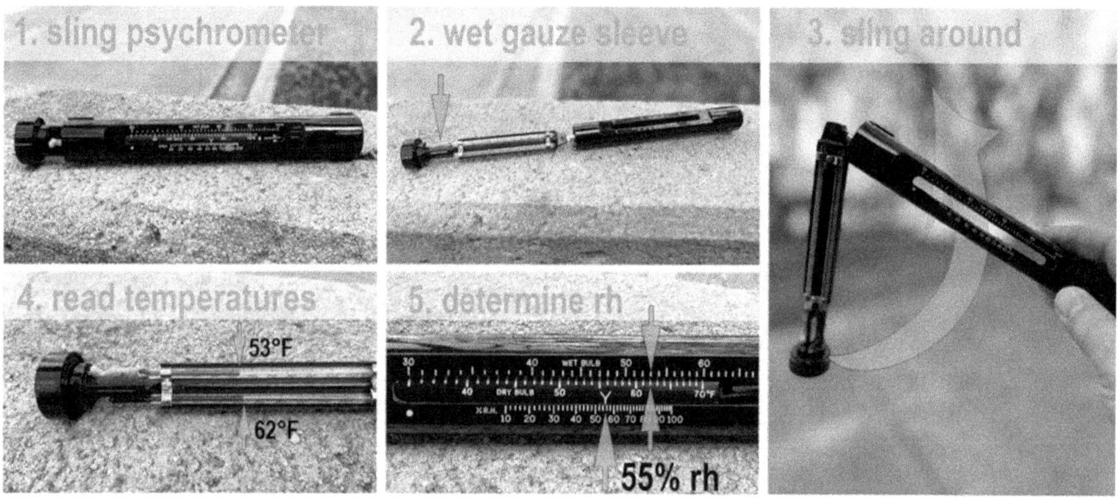

Figure 5.10 A Sling Psycrhometer Is Used to Measure the Wet Bulb and Dry Bulb Temperatures of an Air Sample to Determine Relative Humidity (rh); the Device Has a Built-In Sliding Chart to Align the Wet Bulb and Dry Bulb Temperatures to Determine the Relative Humidity (rh) or We Could Plot the Two Temperature Readings on the Psychrometric Chart; the Device Pulls Apart and Has a Hinged Pivot Point Allow the User to "sling" the Thermometers in a Circular Motion; In This Case, the Dry Bulb Temperature Is 62°F and the Wet Bulb Is 53°F with a Resulting rh of 53%.

In Figure 5.11, a measurement has been taken with the sling psychrometer and the two temperature readings (dry bulb 75°F and wet bulb 64°F) are plotted on the Psychrometric Chart. The **wet bulb temperature** is indicted on the Psychrometric Chart with a diagonal axis. The range of temperatures indicated in the chart goes from 30°F to 90°F with the axis extending a bit further to the upper right-hand corner. From Figures 5.11, we can find the relative humidity (53% rh) by plotting the dry bulb and wet bulb temperatures. From that point, we project to the right along the line for humidity ratio, we can attain that value—0.010 lb. water/lb. dry air.

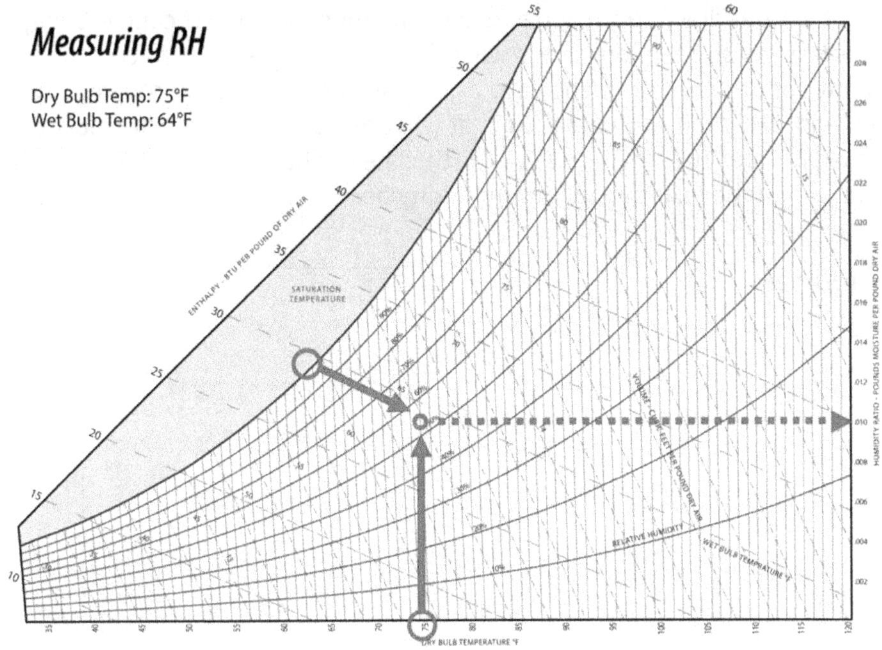

Figure 5.11 The Dry Bulb and Wet Bulb Temperatures Measured with a Sling Psychrometer Are Plotted on the Psychrometric Chart to Determine the Relative Humidity of the Air Sample; 75°F Dry Bulb and 64°F Wet Bulb Yields a Relative Humidity of 53%; 53% Is Generally Considered a Comfortable Relative Humidity Across a Range of Nominal Temperatures from Say 68°F to 78°F.

We have all experienced dew forming in the grass, or on vegetation, or a foggy morning. Or, when we take a can of soda out of the refrigerator, or pour a glass of iced tea, soon the outside surfaces of the glass or can become wet. What happened to cause this? When air is cooled, by the cool surface of the container, its capacity to hold water is reduced. The vapor in the air near the container must condense back Into water, and it needs a surface to form on. The surface might be grass, your soda can, or particulate in the atmosphere that subsequently causes rain or snow. This could also happen in a building wall cavity or on building surfaces if they are colder than the air.

Figure 5.12 A Simple Water Bottle Taken Out of a Refrigerator, Perfectly Dry in the Refrigerator, Becomes Covered in Water Droplets; as the Cold Water Bottle, Approximately 40°F, Cools the Air Around It, the Air Is Cooled to the Dew-Point Temperature, Becomes Saturated, and Additional Cooling Causes the Moisture in the Air to Condense on the Water Bottle.

We can track and predict this process of condensation using the Psychrometric Chart. In Figure 5.13, an air sample is plotted at 75°F and 53% rh and we begin to cool the air. The air might be cooled by coming into contact with a cold building surface or we may be mechanically cooling the air. In any of these cases, condensation will form if the air is cooled beyond the saturation temperature. From the illustration in Figure 5.12, we can see that at approximately 57°F, the line tracking the cooling air intersects with the saturation temperature. When cooled air intersects with this curved axis at the left of the Psychrometric Chart the air must track along the axis and moisture must come out of the air. You might also notice this curve corresponds to 100% relative humidity. At this point in the cooling process, as mapped on the Psychrometric Chart, the air is full in terms of its ability to hold moisture. Continued cooling causes the moisture in the air to condense onto adjacent surfaces.

Saturation is the term used to describe air that is cooled to the point it cannot hold any more moisture. The temperature when this occurs, for a given air sample is the **dew-point temperature**. At temperatures below the dew-point temperature, air must give up moisture to the surfaces nearby in the form of condensing water following along the saturation axis as is shown on the Psychrometric Chart. Notice that as moisture condenses out of the air, the Humidity Ratio that tracks the amount of water in the air, is being reduced in this process. Now there is less moisture in the air and water has condensed on a surface.

Try plotting a few conditions you might have experienced such as a hot muggy day with perhaps a temperature of 90°F and 70% rh. If you cool the air, at what temperature does the moisture begin to condense? If you cool the 90°F and 70% rh air to say 35°F, then heat it back up to 110°F, as it might be in an air heating system, what would the resulting relative humidity be?

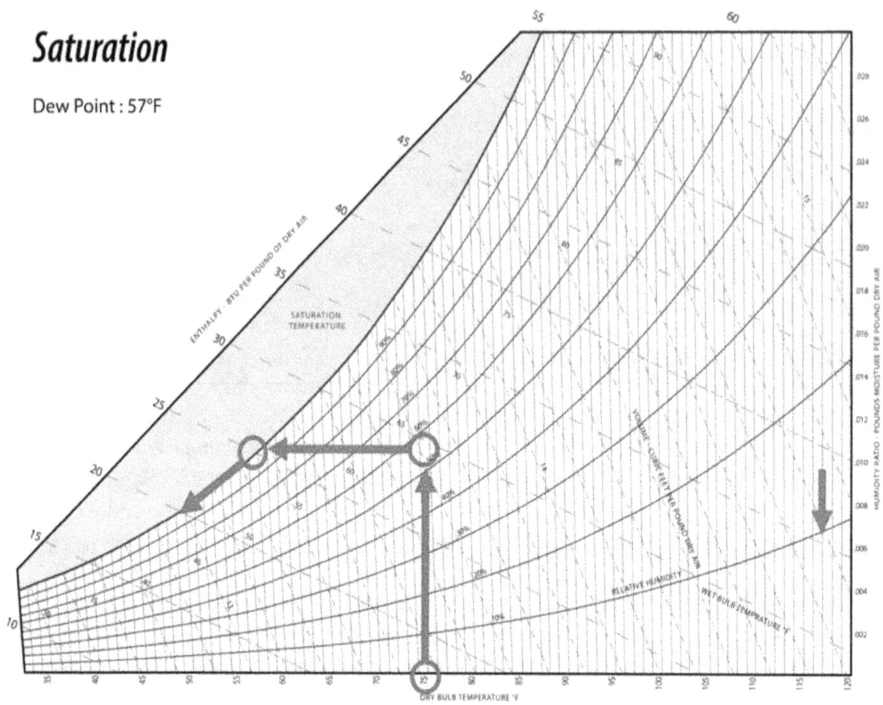

Figure 5.13 Illustrating the Saturation Process as Warm Moist Air Is Cooled Below the Saturation (Dew Point) Temperature, Moisture Must Come Out of the Air; This Moisture Typically Condenses on Surfaces Nearby.

This is a critically important concept for architects because when we cool air, moisture comes out of the air condensing as liquid water on building surfaces. When this happens in buildings, it causes water damage to materials, reduces the capacity of building insulation, and allows molds and funguses to grow on building surfaces. Understanding the relationship between air and moisture is also particularly important for architects seeking to use sustainable design strategies that naturally ventilate, heat with the sun's energy, and employ materials to mitigate external temperature swings.

Enthalpy is another metric that measures moist air and can be tracked on the Psychometric Chart. Enthalpy is the total heat contained in an air sample. This includes the latent heat from moisture and the sensible heat measured by the dry bulb temperature. Enthalpy is measured in BTU per pound of dry air. The enthalpy axis is very similar to the wet bulb temperature axis but is slightly different and the units of enthalpy are indicated at the left and top of the Psychrometric Chart ranging from 10 to 60 BTU/lb dry air.

The situation charted on the Psychrometric Chart in Figure 5.14 with the red dot and arrow shows an air sample that was hot and dry (at the dot) that became cooler and more humid at the point of the arrow. Because the air was quite dry, we can sprinkle or spray water into the air and it will readily evaporate absorbing heat in the process. This process draws sensible heat from the air effectively cooling the air and reducing the sensible temperature. If you check the dry bulb temperature in this scenario, it has been reduced by 15°F. If you check the humidity ratio, you will find there has been 0.008 lb. moisture/lb. dry air added to the air sample increasing from 0.007 to 0.015 lb. moisture/lb. dry air. As this change occurs along a constant line of enthalpy, the "total head content" has not changed. What has changed is the ratio of heat transferring from sensible heat to latent heat as part of the natural evaporation process. We did have to add moisture to get this to occur, but we did not have to add any energy to cool the air. Generally, this new thermal condition would be more comfortable as the humidity is still generally low at 50% rh. In this case of adding moisture to the dry air, we have made the air more comfortable without using any energy to cool the air. This is considered an **adiabatic** process where there is no loss or gain of heat. The enthalpy, or total heat, remains unchanged. This process is also known as evaporative cooling. It is very effective in hot arid climates such as the American Southwest. Many homes in these arid climates have evaporative coolers often referred to as “swamp coolers.” It is important also to understand that if the air is already humid (above 60%) such as in humid climates, then this process becomes much less effective and it probably not viable as a cooling strategy.

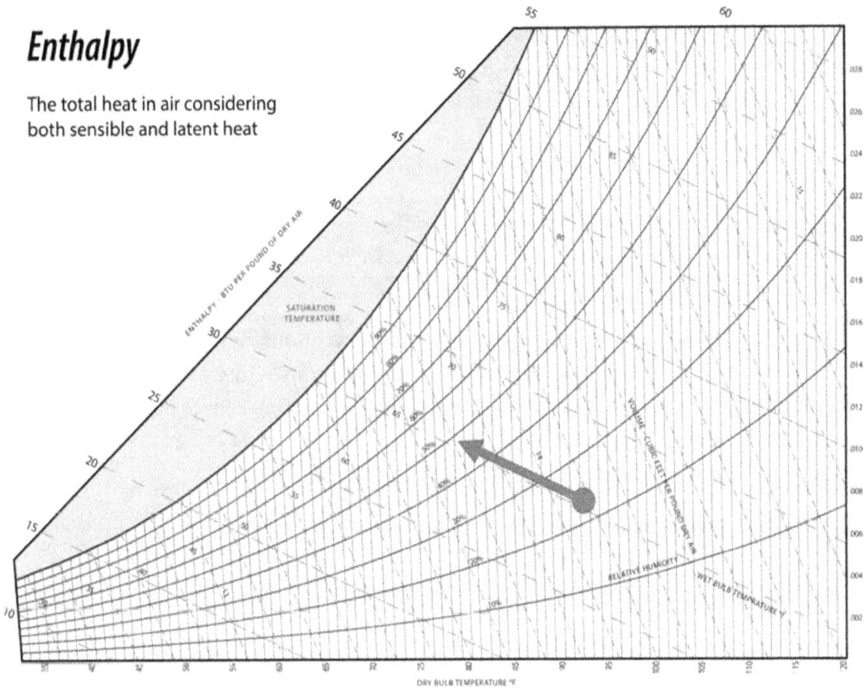

Figure 5.14 Illustration of Air Changing on a Constant Line of Enthalpy With No Change in the Total Heat Content in the Air; One Can See Moisture Is Added as the Humidity Ratio Is Increased, While the Dry Bulb Temperature Has Decreased; This Adiabatic Process Cools the Air Without Using External Energy and Just Adding Water.

Figure 5.15 plots the ASHRAE Comfort Zone on the psychrometric chart. As discussed in the module on thermal comfort, this is a range of thermal conditions in which people will find comfortable considering the conditions established to define the Comfort Zone—the person is male, is dressed in a business suit, and is doing sedentary activity. Of course, this is a narrow range of the people and activities that might occur in air-conditioned spaces. There are adjustments for acclimatization with summer and winter comfort zones. The ASHRAE comfort zone is a good general guide in terms of establishing comfort. However, research and thoughtful judgment is needed to determine the range of temperature and humidity that might be desired for a particular program or use in architectural spaces. With a desired comfort zone, one can evaluate how often those thermal conditions naturally occur at a given location (just open the windows) and also, how much heating or cooling might be needed to maintain comfort.

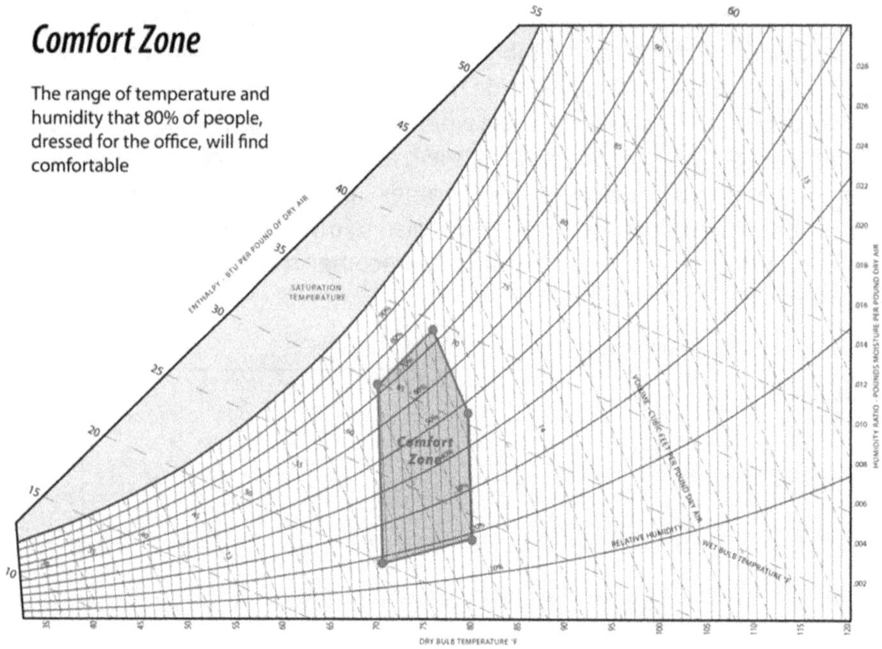

Figure 5.15 The ASHRAE Comfort Zone Is Plotted on the Psychrometric Chart for Summer and Winter Conditions; It Is Important to Note, That This Range of Temperature and Humidity Would Be Comfortable for Middle-Aged Men Wearing Clothes Similar to a Three-Piece Suit and Doing Sedentary Office Types of Activities.

Check out this link for an interactive version of a Psychrometric Chart:

https://demonstrations.wolfram.com/ReadingAPsychrometricChart/

Answers

What would be the relative humidity if we add 0.005 lb. water/lb. dry air to the same air sample? [80% rh]

If you cool the 90°F and 70% rh air to 35°F, then heat it back up to 110°F, as it might be in an air heating system, what the resulting relative humidity would be? [Less than 10% rh.]

References

ASHRAE. (2021). *ASHRAE handbook—Fundamentals (I-P)*. Ashrae.

Bradshaw, V. (1993). *Building control systems* (2nd ed.). Wiley.

Grondzik, W. T., Kwok, A. G. (2019). *Mechanical and electrical equipment for buildings* (13th ed). Wiley.

Lindsey, R. (2009, January 14). *Climate and Earth's Energy Budget*. Nasa.gov; NASA Earth Observatory. https://earthobservatory.nasa.gov/features/EnergyBalance

Image Credits

Fig. 5.5: The American Society of Heating, Refrigerating and Air-Conditioning Engineers (ASHRAE), "ASHRAE Psychrometric Chart". Copyright © 2023 by Techstreet.

Fig. 5.6: Adapted from The American Society of Heating, Refrigerating and Air-Conditioning Engineers (ASHRAE), "Psychrometric Chart – Sensible Heat." Copyright © 2023 by Techstreet.

Fig. 5.7: Adapted from The American Society of Heating, Refrigerating and Air-Conditioning Engineers (ASHRAE), "Psychrometric Chart – Latent Heat." Copyright © 2023 by Techstreet.

Fig. 5.8: Adapted from The American Society of Heating, Refrigerating and Air-Conditioning Engineers (ASHRAE), "Psychrometric Chart – Relative Humidity (rh)." Copyright © 2023 by Techstreet.

Fig. 5.9: Adapted from The American Society of Heating, Refrigerating and Air-Conditioning Engineers (ASHRAE), "Psychrometric Chart – Summary Condition." Copyright © 2023 by Techstreet.

Fig. 5.11: Adapted from The American Society of Heating, Refrigerating and Air-Conditioning Engineers (ASHRAE), "Psychrometric Chart – Measuring RH." Copyright © 2023 by Techstreet.

Fig. 5.13: Adapted from The American Society of Heating, Refrigerating and Air-Conditioning Engineers (ASHRAE), "Psychrometric Chart – Saturation." Copyright © 2023 by Techstreet.

Fig. 5.14: Adapted from The American Society of Heating, Refrigerating and Air-Conditioning Engineers (ASHRAE), "Psychrometric Chart – Enthalpy." Copyright © 2023 by Techstreet.

Fig. 5.15: Adapted from The American Society of Heating, Refrigerating and Air-Conditioning Engineers (ASHRAE), "Psychrometric Chart – Comfort Zone." Copyright © 2023 by Techstreet.

Module 6

Thermal Properties and Heat Flow

Martin Gold

If you can't (under)stand the heat—get to know it a little better.

Introduction

In this module, we will discuss the mechanisms of heat flow that effect buildings and that interact with building materials. In the module on psychrometrics, we focused on the interchange of heat between water and air—humidity—and how the humidity (latent heat) and temperature (sensible heat) within the air might affect your thermal comfort. In this module, we will focus on air, air movement, and how heat flows within and between air and architectural materials. If you are doing this module ahead of the module on psychrometrics, not to worry, the information will still be relevant and students can do the psychrometric module later.

Five Factors of Heat Flow

1. Sensible heat (air temperature)
2. Latent heat (moisture in the air)
3. **Convection (air movement)**
4. **Conduction (objects in contact)**
5. **Radiation (objects in view)**

Bolded topics are covered in this module. Factors 1 and 2 are covered in the module on psychrometrics.

Figure 6.1 Mission San José y San Miguel de Aguayo; the Deep Loggia Shades the Building While the Massive Limestone Building Material Balances the Day and Balance Heat Gain in the Day with Heat Loss at Night to Provide Thermal Comfort in the Mission.

Thermal Environment

Mission San Jóse (Figure 6.1 below) captures a dramatic moment on a hot October day just outside of San Antonio, Texas. There is a sharp contrast between the shadows and bright sunlight with the clear blue sky beyond. The air is warm, and the sun, although low in the sky this time of year, is still quite intense. We would be seeking the shade to be comfortable even in light clothing. The deep archway, originally covered with a wood roof system, will shade the main spaces from the sun during the long and hot summers. However, in the evening, after the sun sets, this dry climate can be quite cool and perhaps even chilly. During the evening, the heat remaining in the mission walls, from the sun shining on them during the day, is given back to the cool night air and also radiated to the dark and cold night sky. Even in the daytime, as the sun warms the ground and walls of mission, heat is also radiated to the sky or more precisely to outer space. This daily cycle of solar heating and night cooling, along with the shading and thermal mass of the walls, keeps the interior spaces comfortable. Of course, there is no modern air conditioning in the mission. The thermal mass, as a thermal strategy, is one way to provide comfort during the summer. A fire can warm the space in the winter. The early Spanish and Mexican settlers understood how the sun transferred heat to the building, and how the building could use this heat effectively. They selected heavy materials with high thermal mass available in the local area to provide thermal comfort inside the spaces even when the outside temperatures were too hot or too cold.

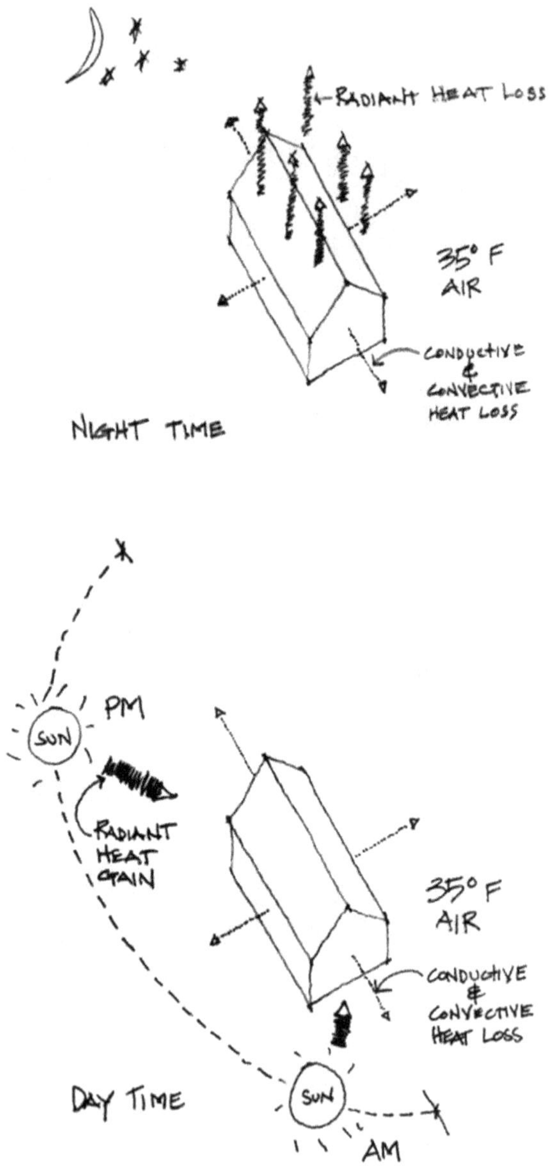

Figure 6.2 Diagrams Illustrating the Impact of Both Heat Loss, and Heat Gain in the Scenarios Associated with Steady State and Periodic Heat Flow in Architectural Systems (Proof of Concept Sketch Currently in Place—Needs Refinement).

Heat Flow

When we are trying to understand the principles of heat flow and heat exchange between and within building materials, it is critically important to keep in mind that heat moves naturally from areas of high concentration to low concentration. ***Heat moves from hot to cold***. When we are discussing heat, we are always following the heat. For example, when we are cooling spaces, we should conceptually be thinking about removing the heat. And, of course, we should understand where we are putting the heat when we remove it—it must go somewhere.

Architects are typically concerned with two types of heat flow: steady state and periodic. This will be discussed in more detail later in the module, but let's introduce the concepts here. **Steady state heat flow** occurs when there are two, generally stable temperature regions, and the heat flow (moving from hot to cold) is constant. For example, on a cold winter night, the outside temperature might stabilize early in the morning (5:00 am) at say, 38°F and you may have your inside thermostat set at 68°F—a temperature difference of 30°F. This steady state condition makes it easy to calculate the heat flow because it is driven by the simple temperature difference—30°F. Alternatively, **periodic heat flow** typically occurs during the day, when the sun is shining on a building. In this case, the surface temperature of the building material could rise to between 130°F and 180°F, while the air temperature could be fairly constant at temperatures as low as 65°F. As the sun moves off the surface, the material will start to cool giving up heat to the air until it matches the air temperature. This is a dynamic situation that requires information about the material's mass, reflectivity, and heat capacity to make predictions regarding how heat might be flowing into or out of a building through the envelope.

When we measure heat flow, we typically measure in British thermal units (Btu) at an hourly rate (Btu/hr). For reference, 1 Btu is equal to the amount of heat needed to raise the temperature of 1 pound of liquid water from 39°F to 40°F (1°F). Interestingly, 39°F was chosen for this reference as 39°F is the temperature where water has its greatest density.

Convection

Few things are more awe-inspiring than a brightly colored hot air balloon floating through the atmosphere. The beauty is even more palpable knowing the simplicity of the principle—hot air rises as cool air sinks. The sinking cool air, being heavier, actually pushes the lighter less dense air in the balloon upward which is the physical mechanism of convection. The hotter the air, relative to the surrounding air, the faster and higher the balloon will rise. However, as the balloon continues to rise, the atmosphere itself becomes less dense at higher altitudes and, even though high altitude air gets much colder relative to the balloon air, the balloon will no longer rise—a natural protection that keeps balloons from floating off into space.

Figure 6.3 Hot Air Balloon, Illustrating the Power of Thermal Convection—Warmer Air Naturally Rising Within Cooler Air Because Warm Air Is Less Dense Than Cooler Air.

Convection occurs in all fluid media such as water or air. Gases and liquids are considered fluid media. When a portion of the fluid—let us use air in this case—is heated, the molecules become more energetic pushing harder on each other when they collide and subsequently the volume of air expands, becomes less dense relative to the surrounding air, and starts to rise. The warmer, less dense volume of air rises because gravity pulls the denser cooler air down filling in the space below the warmer air pushing it away from the ground. This gravity-driven phenomenon consistently pulls cooler air down allowing warmer air to rise.

If we apply this concept to the architectural situation of warming a space with sunlight in winter, we can track the convective movements of air within a space. Figure 6.4 illustrates heat flows as sunlight comes through a window to warm surfaces in the space and subsequently, warms the air adjacent to those surfaces. This warm air will rise. The outside air is cold and therefore so is the window. The inside air near that window will be cooled by the window surface and it will fall as it becomes more dense and is pulled down by gravity. This rising and falling of warm and cold air can initiate a cycle, or a **convective loop** that can draw the heat out of the space by transmitting it to the exterior through with window. This convective loop phenomenon can also occur in wall cavities, attics, and other void spaces in building systems. One of the pri-

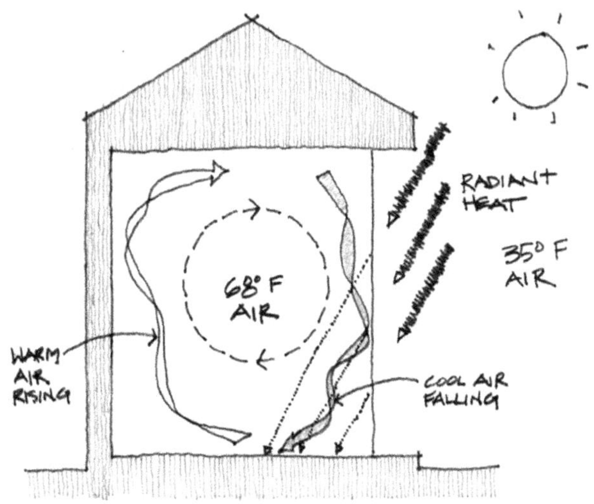

Figure 6.4 Sketch Illustrating How a Convective Loop Can Form in a Space and Increase Heat Loss During Cold Winter Months; the Sun's Radiant Heat Energy Transmits Through the Window, and Then Warms the Floor Slab Inside the Space; Then the Warm Slab Transmits That Heat to the Air by Contact.

mary important actions of building insulation is to prevent convective loops from occurring in wall cavities and attics.

Convection can also lead to **stratification** of the air in a space. In this case, warm air pools at the top of a space near the ceiling while the cooler air pools closer to the floor—where people occupy the space. If convective loops do not form, the warmer air will stay at the top of a room. This can be desirable in hot and warm climates. Rooms with high ceilings can allow warmer air to pool at the top of the space adding an insulation layer from a warm ceiling. Rising warm air can be drawn up and out through clerestory windows subsequently drawing fresh and cool air into the lower space as a passive ventilation strategy. Passive systems will be discussed later in future modules. The concept of convective air flow is fundamentally important to understand with respect to applying passive design strategies.

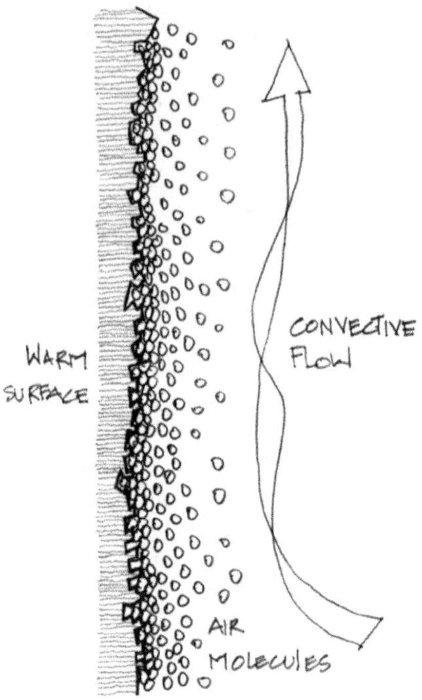

Figure 6.5 Molecular Interaction of Air and Building Surfaces Illustrating How Molecules Are Trapped at the Surface and How Both Natural Convection and Wind Can Disrupt This Insulating Air Layer.

At the molecular level, air seeks to cling to building surfaces as molecules of air can become trapped in the voids and by the mechanism of capillary action. This air film that builds up on the wall provides a small insulating layer that reduces heat flow. **Air is the best insulator**. We will revisit this concept in other modules but let's introduce it here while we are considering air at the molecular scale. Air is the least dense commonly occurring material in buildings. When we use "insulation" as a material, that material is just holding the air in place. It prevents convective loops from forming in wall cavities. It is not the insulation itself that "insulates," it is the air that is trapped by the insulating material. When the thin layers of air at within walls are get hot or cold, convection takes the insulating layer away from the wall surface reducing the insulating capacity of the air film. Of course, wind on the exterior of a building does the same thing at the exterior—removes the molecular air film that is insulating the building.

Conduction

Conductive heat flow (conduction) occurs within materials and between materials that are in direct contact with each other. In this case, the molecular energy within molecules is transferred directly by bumping into other molecules—also referred to as kinetic exchange. This physical contact between molecules allows heat to move through a single material and to transfer heat from one material to another so long as the materials are in direct contact. For example, if you put a pot of water on an electric range top burner and turn on the burner, the heat is transferred within the burner by conduction, then transferred to the pot on the stove by conduction through contact, then through the material that makes up the pot by conduction, and then to the water in the pot that you may be trying to boil via conduction. Heat transfer within the boiling water as it moves within the pot, also occurs through convective loops that are visibly evident by the rapid movement of the boiling water. You might hover your hand over the burner and feel the heat of the burner (not touching it of course), but this heat you feel is from radiation because you are not touching it. Radiation is discussed below. For conduction to occur between objects, they must be in contact with each other.

When heat flows through a simple (homogeneous) material—having the same consistency all the way through—the rate of heat flow (heat flow per hour), is driven by the temperature differences on each side of the material. The greater the temper difference, the greater the heat flow. For example, if the temperature on either side of the material is the same, let's say, 75°F (75°F outside and 75°F inside) then the difference between those temperatures (75°F - 75°F = 0°F) is zero. We would say that the "delta t" (Δt), the difference, between the temperatures at the inside and outside surfaces of the material is zero. In this unique case, no heat would flow. Figure 6.6 illustrates a condition where the outside air temperature is 95°F and the inside air temperature is 75°F so the difference (Δt) is 20°F. Therefore, heat is flowing from hot 95°F to the cooler 75°F through the building material. Heat always flows from hot to cold.

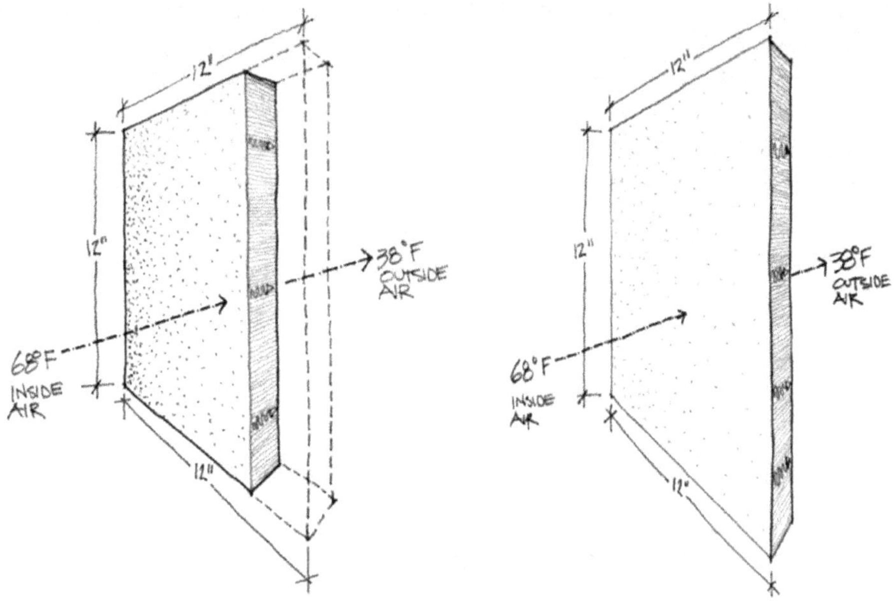

Figure 6.6 Conductivity of a Simple Building Material; This Section Cut Perspective Shows Heat Flowing Through a 1¢ × 1¢ ($12^2 \times 12^2$) Square of Building Material; Heat Transfers at the Surface by Convection/Conduction and Then From One Surface Through the Material to the Other Surface by Conduction; Then to the Outside Air by Conduction/Convection; Conductivity Is a Measure of How Much Heat Transfers Over a Given Time Period Through 1 ft^2 of Building a Material.

The next question one might ask Is: how do different materials either transmit or resist transmitting heat? This depends on the properties or "materiality" of the building material. There are four common metrics used to rate the amount of heat flow that occurs through a variety of materials. We may want to know how good a material is at conducting heat, or we may want to know how good a material is at resisting heat flow or insulating against the flow of heat. The following metrics are commonly used depending on the information needed for a specific design question. The metrics listed below measure the heat flow through a 1 ft^2 area of material, as illustrated in the diagram in Figure 6.6. In an actual building situation, the heat flow would be multiplied by the total area (in square feet) of the material being used in the building. We will review the application of these principles in later modules.

Measures of Conductive Heat Flow

Conductivity (k) is a measure of the heat flow through a *homogeneous* material per inch of thickness with units: Btu/(hr·ft^2·°F). For example, if you have a homogeneous material such as concrete, that is 3 inches thick, you would multiply the "k" value by 3 (based on the 3 in. thickness) to get the conductivity of that 3 inch thick material.

Conductance (C) is a measure of the heat flow through a material that is *non-homogeneous* and for a given thickness with units: Btu/(hr·ft^2·°F). In this case, materials such as concrete masonry units (with both solids and air voids within them) or insulating materials (with air spaces) are not consistent with regard to heat flow as they get thicker or thinner. Therefore, these materials have *conductance* ratings rather than *conductivity* ratings. The conductance ratings are for the thicknesses listed by manufacturers and cannot be reliably translated to other thicknesses.

R-value (R) measures a particular material's resistance to heat flow. Resistance is the inverse of conduction with units: (°F ft^2 hr/Btu). The R-value indicates how well a material prevents heat from flowing through it. Insulating materials are typically evaluated by comparing their R-values. For example, a good glass fiber batting insulation located in a wall cavity of 3-1/2² depth would have an R-value of approximately 11.

U-value (U) is a measure of how much heat flows through a wall assembly that is made up of multiple components with units: Btu/(hr·ft^2·°F). The U-value is the inverse of the sum total of all the individual material R-values in a wall assembly. This is an important metric and is used to calculate how much heat is flowing through a given wall assembly. This is the primary thermal unit used to size mechanical systems.

In practice, one would find the R-value for each building element in a wall assembly that could have perhaps 1 to 10 components, then add up the R-values to get the total (the sum). Then take the inverse of that (1/R-value total = U-value) to find the U-value of the assembly. We will revisit this concept and its application in later modules.

Formula for calculating the U-value for a wall assembly.

$$U = \frac{1}{\sum R_{1-N}}$$

Add R_1 R_2 R_3... R_N, to get the sum ($\sum$) of the R-values for each material.

Steady State Heat Flow Revisited

Let us consider an early morning winter condition. The sun has set long ago, and the nights are long during the winter. In this example, the outside temperature has generally stabilized at say 32°F and the inside thermostat is set at 68°F—the temperature we are trying to maintain by heating the building. 32°F is also the lowest temperature generally occurring for the location of this building (southern coastal US). This data we can find online and will discuss in more detail in other modules on climate. So, let's use 32°F as the ***outside design temperature*** for determining how much heat is being lost (heat loss). If we know the thermal resistance (U-value) of the walls, roof, and windows, and the areas of each of those building assemblies, in addition to the outside design temperature, it is then simple to calculate how much heat is being lost to the outside. With that information we can determine how large our heating system needs to be. Subsequently, we can estimate how much energy we can expect to use to heat the building. Current computer models do this calculation for every hour of the year—all 8,760 hours. This steady state situation is simple because we only need to know the outside air temperature (°F), inside desired temperature (°F), the thermal transmittance (U-value) of the building assemblies, and the area (A) of the building assemblies. This relationship can be shown in the formula below.

Formula for calculating the heat flow through a given building assembly—walls, windows, roof systems.

$$Q = U \times A \times \Delta t$$

Q = heat measured in Btu/hr

U = U-value of the building assembly in Btu/(hr·ft·°F)

A = the area of the building assembly in ft^2

Δt = the difference in temperature across the surface ($t_{inside} - t_{outside} = \Delta t$) in °F.

This formula represents fundamental heat flow relationships architects should well understand to be able to evaluate and select materials and assess thermally related design decisions.

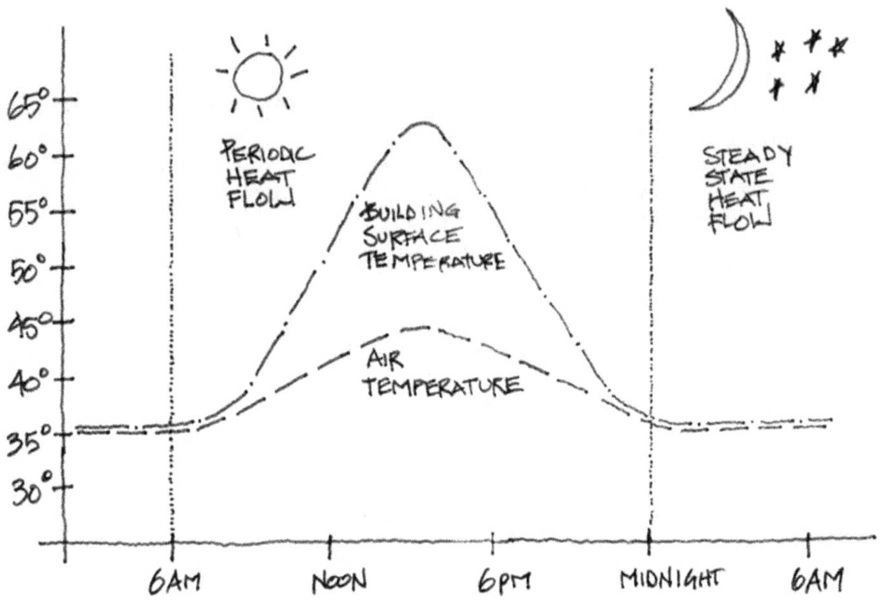

Figure 6.7 Steady State Heat Flow Occurs at Night, After the Sun Sets and Both Are and Surface Temperatures Have Stabilized; Periodic Heat Flow Occurs During the Daytime, When the Sun Rises Imparting Radiant Heat Energy to the Surfaces of Buildings and the Surrounding Areas; As the Sun Transits the Sky, Building Surfaces Facing East, South, West, and North and the Roof Will Be Affected Differently Throughout the Day.

Radiant Heat Flow

Radiant heat flow travels by means of electromagnetic waves between objects over distances that have a line of sight between them. The objects are not physically touching each other. In the case of the sun, radiant heat travels to the earth over a vast distance of 94 million miles and is filtered as it travels through earth's atmosphere. We sense heat from the electromagnetic spectrum in the frequency range of approximately 10^{11} to 10^{17} Hz (wave cycles per second). Within the range of heat, we can "see" the smaller range of electromagnetic waves that we perceive as light—as illustrated in the image below. There are many other categories of electromagnetic waves ranging in wavelength from gamma rays to long wave radio waves. For architects, it is important to understand ultraviolet (UV) short wave radiation (wavelengths around of 10^{-8} m) and infrared radiation (IR) long wave radiation (around 10^{-6} m to 10^{-4} m). Please note that as the frequency (n) in Hertz increases, the wavelength (l) decreases having an inverse proportional relationship.

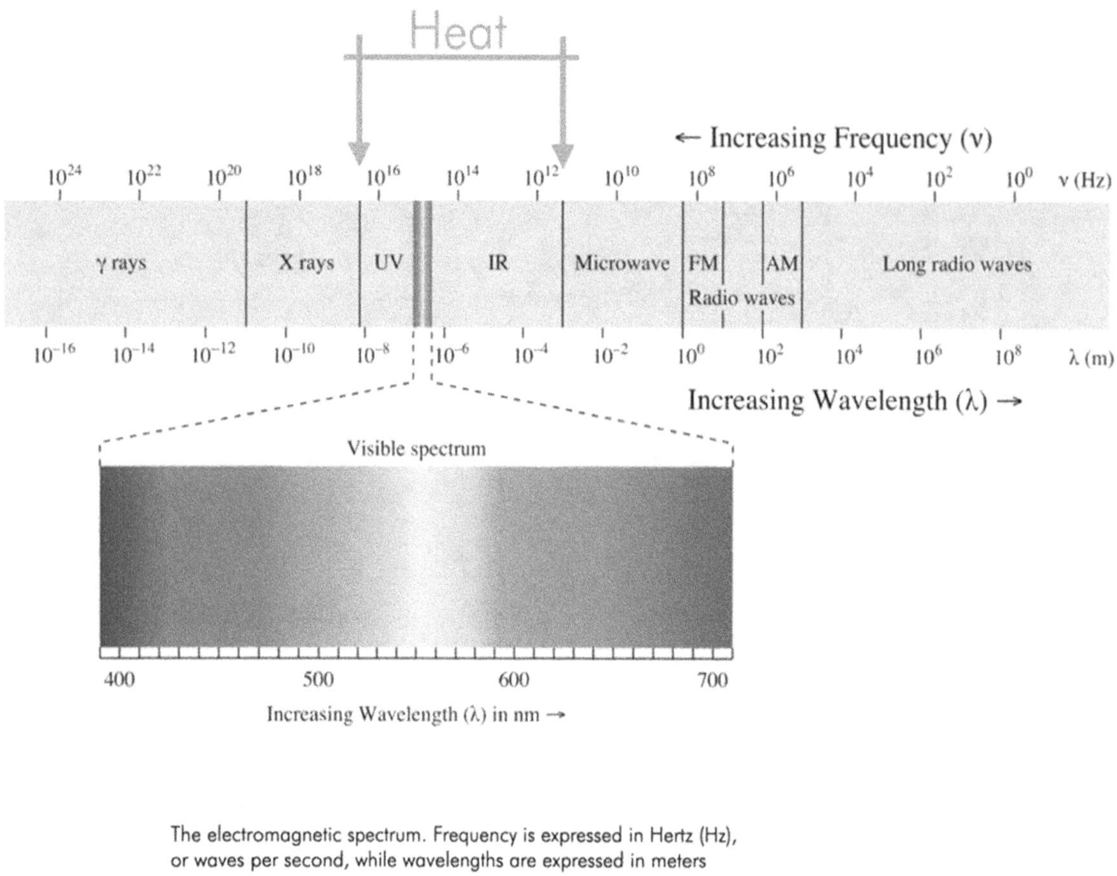

Figure 6.8 Electromagnetic Spectrum Showing the Range of Frequencies of Ultraviolet (UV) Through Infrared Radiation (IR) That Are Perceived as Heat; Note That Ultraviolet Light Has a Wavelength of Approximately

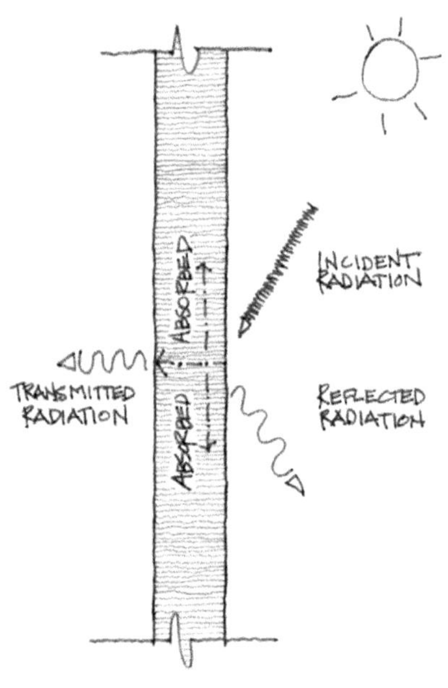

Figure 6.9 Sketch Diagramming the Elements of Radiant Heat Exchange with a Simple Building Material.

Radiant heat exchange, like conductive heat exchange, is driven by the *temperature difference (Δt)* between two or more objects. Because radiant heat travels over distance, and interacts with objects through their surface qualities¾how reflective or absorptive the surfaces are¾the material properties are the primary factor in how much radiant heat exchange takes place between objects. Radiant heat exchange is somewhat moderated by the reflectivity, color, and mass of the material. Radiant heat exchange can be quite complex.

As illustrated in the sketch in Figure 6.9, when radiant energy strikes a surface, some portion is reflected, some portion is absorbed (warming the material), and some portion is transmitted through the material. The heat can be transmitted through radiation, conduction, and convection as the warm surface interacts with the air in the space. In addition, the angle at which the radiant energy arrives at the surface changes the amount of radiant heat flow. For example, radiant energy arriving perpendicular to the surface will transmit more heat energy than if it arrives at an angle such as 45°.

When considering radiant energy, the temperatures of objects can be quite high or low with extreme temperature differences. For example, a bonfire or fireplace can be 2,000°F and the clear sky can have temperatures as low as 8°F. In the illustration in Figure 6.10, the ground temperature is shown to be 68°F. Typically, ground temperature does not vary much from day to night as its mass is high—it will vary seasonally. However, on a hot summer day, in the direct sun, pavement and asphalt surfaces can heat to temperatures over 130°F and lighter weight materials (less mass) can be heated to over 180°F by solar radiation. As heat moves from hot to cool, we can observe that when the sun rises and starts heating the environment, we are exposed to multiple varied surface temperatures that are typically rising throughout the day and then falling in the evening.

Delta ts (Δt) for the various heat exchanges shown in Figure 6.10.

Fire to sky (1880°F – 8°F): Δt = 1872°F

Fire to ground (1880°F – 68°F): Δt = 1812°F

Fire to skin (1880°F – 97°F): Δt = 1783°F

Skin to sky (97°F – 8°F): Δt = 89°F

Skin to ground (97°F – 68°F): Δt=29°F

Figure 6.10 Image Illustrating the Various Radiant Heat Exchanges Between People and Their Environment in the Case of a Bonfire on a Cold Night.

The *relative size of objects* that are exchanging heat is another important factor in how much radiant heat will flow. Although the fire (Figure 6.10) may be very hot and the sky very cold, they offer different surface areas relative to their positions to us. For example, as we get closer to the fire, it feels warmer even though the temperature of the fire has not changed. To understand the relationship between the radiant heat transfer and "visual size" of the object, we consider the **subtended solid angle of view.** If we were to imagine a cone from our line of site to the perimeter of an object or surface in our proximity, that cone represents the solid angle of view. This angle can be used to measure the radiant impact of that heat source. Consider the illustration below with a single campfire and two campers, one with a 20°F solid angle of view and one with a 40°F solid angle of view. The closer we get to the object, in this case the campfire, the 'larger it appears to be' and subsequently the larger the solid angle of view. And of course, the closer we are to the fire, the warmer it feels.

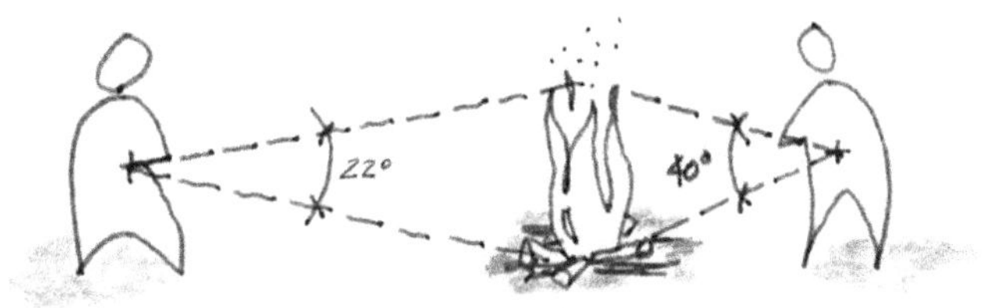

Figure 6.11 Sketch Illustrating the Visual Size of an Object with Respect to the Amount of Radiant Heat That Might Be Transmitted.

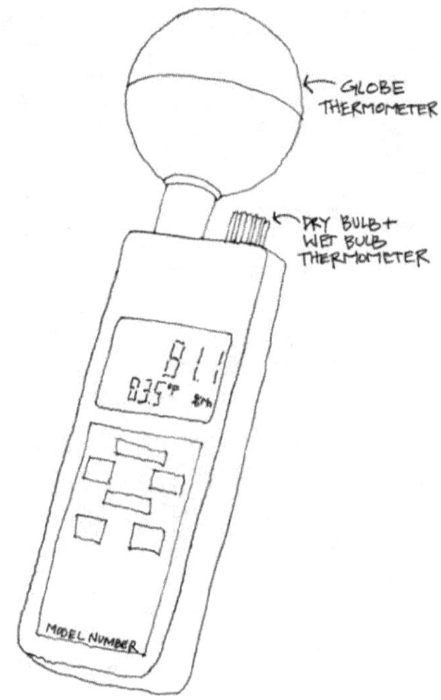

Figure 6.12 Globe Thermometer's, With a Large Black or Gray Sphere, Can Be Used to Measure the Cumulative Radiant Impact of a Given Space; Illustrated Here Is a Modern Thermometer Containing a Globe Thermometer (Large Sphere), a Dry-Bulb, and a Wet-Bulb Thermometer.

Mean radiant temperature (MRT) is a measure of the overall impact of the radiant heat flow from the objects that might be around us. We could measure MRT with a globe thermometer, a thermometer that is essentially a regular thermometer located inside an enclosed sphere of lightweight material painted black or dark grey. The black sphere accounts for the influence of radiant heat exchange in addition to the ambient air temperature. Often, these temperatures are combined with a wet-bulb thermometer to give readings of air temperature, relative humidity, and mean radiant temperature. Single devices can measure radiant heat, latent heat, and the sensible temperature for use in calculating the heat index or cumulative effects of these measures on comfort. These are becoming commonly used tools at outdoor sports training facilities to assess the potential health risks to athletes of heat stroke or dehydration. The are also great tools for evaluating potential comfort in buildings.

One could also estimate the mean radiant temperature for a given situation such as that depicted in Figure 6.13 below. In this example, two people (person A and person B) are occupying the same space at two different locations. We can gain a general idea of the surface temperatures from the air temperatures—so long as the sun is not radiating to the outside surfaces. The exterior walls and windows will have surface temperatures between the outside and inside air temperature. The interior partitions will typically have temperatures the same as the inside room air temperature. To calculate the MRT, we simply multiply the surface temperature in °F by the angle of exposure in angular degrees (totaling 360°) and then divide by 360°. This gives us an arithmetic mean (or average) of the influence of the various surface temperatures around us on our perceived thermal comfort.

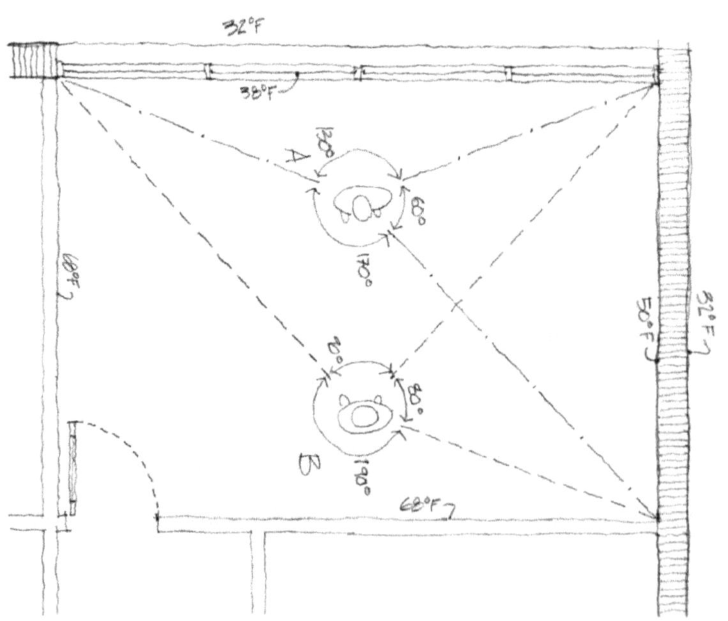

Figure 6.13 Sketch Scenario of a Space Were the Mean Radiant Temperature (MRT) Is Different for the Two People Who Occupy the Room; in This Case, the Sun Is Not Shining on Either the Exterior Window or Wall.

Example. The following simple calculations assess the mean radiant temperature (MRT) affecting persons A and B in the scenario presented in Figure 6.13 above.

Person A: [(130° × 38°F) (60° × 50°F) (170° × 68°F)] ÷ 360° = MRT

[(4,940°°F) (3,500°°F) (11,560°°F)] ÷ 360° = MRT

20,000°°F ÷ 360° = MRT

55.5°F = MRT

Person B: [(90° × 38°F) (80° × 50°F) (190° × 68°F)] ÷ 360° = MRT

[(3,420°°F) (4,000°°F) (12,920°°F)] ÷ 360° = MRT

20,000°°F ÷ 360° = MRT

56.5°F = MRT

Based on this process, we can conclude that person B is slightly warmer than person A. Take a moment to calculate the MRT of person A, given the same outside temperatures, but considering direct sun striking the glass and warming it to 75°F. All you need to do is replace the 38°F value with 75°F for the window surface. What MRT did you get? If you got almost 69°F that is correct—slightly warmer than the ambient 68°F room temperature. Hopefully, you are gaining an understanding of the importance of radiant heat and MRT on thermal comfort. Thermal comfort is much more complex than just understanding how the air temperature affects comfort.

Emissivity is the ability of a *source* surface or object to emit electromagnetic waves. We consider emissivity when we are evaluating how well a surface can emit radiation. If we are trying to maintain heat, we would look for surfaces that are poor emitters. If we are trying to shed heat, we would look for materials that are good emitters.

We need to take a moment here to introduce the concept of a "black body." A black body is a perfect emitter or absorber of electromagnetic waves. Although they are theoretical and nonexistent, the concept is used for the calculations that follow. There are materials that approach being perfect black body absorbers or radiators such as graphite and lamp black. The "black body" concept is used to determine a ratio that compares the material we might be interested in with the perfect black body emissivity.

Materials with high emissivity, close to 1.0, are good emitters. If they have emissivity values closer to 0.0, they are poor at emitting radiant energy to other objects. From the examples listed below, you can see that commonly used building materials are quite good at emitting radiant energy—with the exception of aluminum.

Emissivity (ε) = the emissivity of the source ÷ emissivity of a black body

Example Values of Emissivity

Aluminum	0.03
Asphalt	0.88
Brick	0.90
Concrete	0.91
Glass	0.95
Paint (white)	0.90

Absorptance is the ability of a *receiver* surface or object to absorb electromagnetic waves. In the case of absorptance, we measure the amount of radiant energy absorbed as a ratio of how much energy it received. We could also say, the amount of radiant energy that is not reflected. Black holes are an example of a near perfect black body absorber as they absorb almost all of the electromagnetic energy that strikes them. Here again, values close to 1.0 are good absorbers of radiant energy and values closer to 0.0 are poor absorbers of radiant energy.

Absorptance (A) = radiant energy absorbed ÷ radiant energy received

Example Values of Emissivity and Absorptance

	Emissivity	Absorptance
Aluminum	0.03	0.09
Asphalt	0.88	0.90
Brick	0.90	0.65
Concrete	0.91	0.60
Glass	0.95	0.10
Paint (white)	0.90	0.16

You may find it interesting that some materials are quite similar in absorptance and emissivity (asphalt) and some are quite different (glass and white paint). Perhaps there are situations where these properties can become quite useful in terms of providing comfort and saving energy. One can easily search the internet for the specific emissivity and absorptance of a wide range of building materials and use that information quite intelligently armed with a solid understanding these basic principles.

Radiant heat and the subsequent wavelengths on which it is transmitted offers a particularly fascinating phenomenon with regard to transparent or translucent materials such as gasses and liquids—they do not transmit all electromagnetic wavelengths equally. In other words, some electromagnetic waves (UV) can pass through earth's atmosphere or the types of glass used in buildings—yet others such as IR cannot. If we can see light through a window pane then it is passing electromagnetic waves in the visible spectrum.

We can tune glass to regulate UV wavelengths or even particular wavelengths of visible light—tinted glass, smoked glass, or mirrored glass have these light filtering effects. If we want to warm spaces with sunlight, then we do want the ultraviolet wavelengths to enter and to heat the interior surfaces of the building.

When UV light from the sun enters through a window, and is then absorbed by the surfaces such as the floor and perhaps the walls, the heat is re-radiated as infrared (IR) also referred to as terrestrial radiation. Infrared radiation, however, cannot pass through the glass. In spaces, it is reflected back into the space keeping that radiant heat from escaping. This phenomenon explains why the inside of a car gets much hotter than the outside air temperature on a sunny day. This is the same characteristic of the atmosphere that traps IR radiant heat on earth. UV radiation from the sun can pass through the carbon dioxide and methane in our atmosphere but IR or terrestrial radiation cannot as easily pass back through these gasses to escape back to outer space. This radiant heat is subsequently trapped in the atmosphere. As those IR trapping gasses increase more heat is trapped contributing to the phenomenon of global warming.

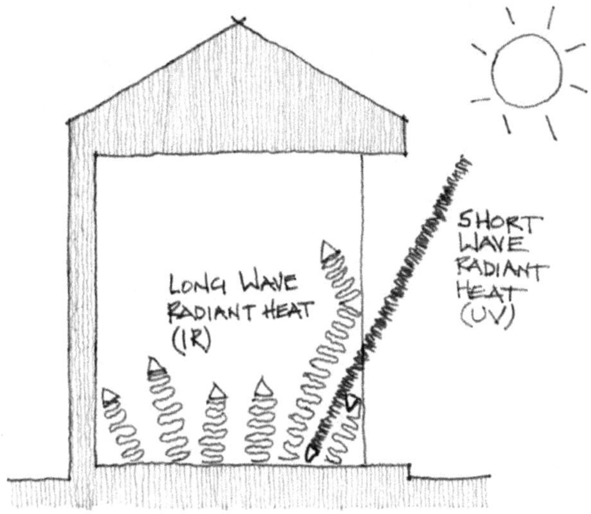

Figure 6.14 The Green House Effect as Applied to a Simple Solar Warmed Space. The Section Diagram Illustrates the Basic Phenomenon Driving Passive Heating With Direct Solar Gain in a Living Space.

The sketch below illustrates how the UV radiation (short wavelengths) can pass through glass (or atmospheric gasses) but IR radiation (long wavelengths) cannot escape and are trapped in the space. If you are trying to warm a space in winter, this natural quality of glass is of great benefit. If you are trying to keep radiant heat out of a space, say in a hot humid climate, you might consider low emissivity glass that prevents much of the UV radiation from entering a space. Here again, understanding material interaction with radiant heat flow can make your architecture much more effective in terms of human comfort and limiting the use of external energy sources.

Periodic Heat Flow Revisited

Equipped with our newly refreshed understanding of conductive and radiant heat flow, we can discuss the implications of periodic heat flow relative to architectural design and material choices appropriate to climate. When the sun rises in the morning and transits the sky throughout the day, it imparts substantial radiant heat to the ground, vegetation, surface water, and building surfaces raising those surface temperatures. Depending on the type of surface, temperatures could reach from 130°F to 180°F.

For example, the sketch in Figure 6.15 below shows that as the sun comes up, it begins to heat building materials—represented as a concrete masonry unit. Because the material is dense and heavy having substantial mass, it takes time to conduct the heat absorbed at the surface through the thickness of the material. Eventually, the heat is transmitted to the interior side and occupied space. Late in the day, the exterior surfaces could get quite hot, but the interior surface will remain cool if the material is thick and dense. The base line in Figure 6.15 represents sunrise and sunset. As the sun is setting, building materials are still transmitting heat from the day to the inside of the building. Only after sunset does the exterior surface of the material start to cool again. At night, heat flows from the material surface through conduction and by radiating to the cold night sky. The heat flow reverses direction from day to night. As all materials have some thermal mass, and some have high thermal mass, it takes for the heat to move through the material. The more massive the material, the longer it takes for heat to transmit through the material. This natural process of heat flow results in two important phenomena: thermal time lag; and the thermal moderating effect.

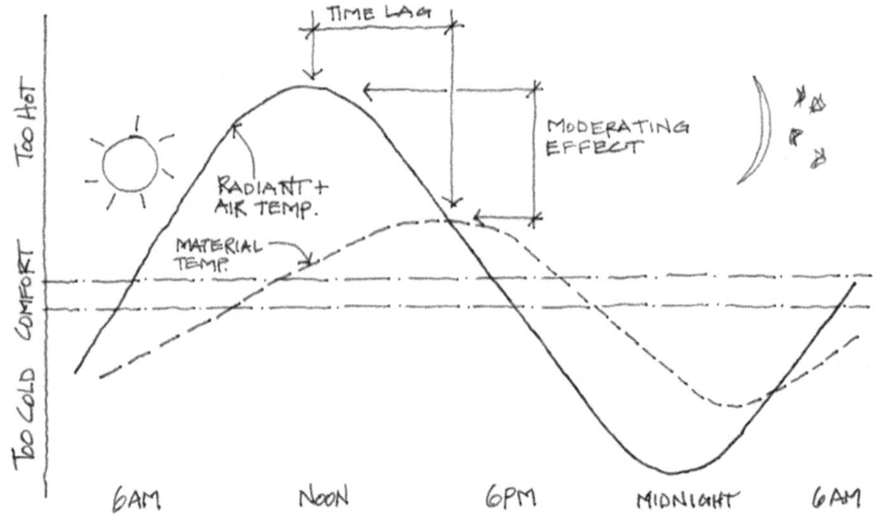

Figure 6.15 Thermal Time Lag and Moderating Effect as a Diagram of the Temperature Fluctuations From Day Into Night and Back Into Day.

Thermal time lag occurs because the peak temperature at the interior surface of the building material happens much later in time than the peak of solar radiant heat strikes the exterior side of the material. If we could tune the material, using the right mass and orientation, to have a peak interior temperature that corresponds to the lowest outside temperature, we can keep a much more even temperature inside a building.

Thermal moderating effect is related to the time lag noted above, but in this case, we are specifically interested in the range of inside surface temperatures that the material experiences relative to the peak radiant temperature at the outside surface. So, even if the outside of the concrete block surface gets quite warm, the inside can stay relatively cool, due to the material's thermal mass. If the sun is only striking and raising the outside surface temperature of the building material for approximately half or less of the 24-hr day and nights are cool, the interior surface will always be moderated by the materials thermal mass having much less extreme temperatures.

These diurnal strategies that result from the periodic radiant solar heat are best applied at latitudes below approximately 60°, and in areas that have high daily temperature swings such as climates like deserts with hot days and cold nights. They can also be adapted to more moderate climates and integrated with other low-energy or passive thermal control strategies.

Summary

Within the previous module and this module, we've considered the five factors of heat flow and given some practical applications of how these concepts can be utilized by architects during design for improved comfort and energy efficiency. Readers should now have a basic working understanding of the following aspects of how heat interacts within our environment and the materials typically utilized in buildings.

Five Factors of Heat Flow

- Air temperature (sensible heat)
- Humidity (latent heat)
- Air movement (convection)
- Objects in contact (conduction)
- Objects in view (radiation)

References

ASHRAE. (2021). *ASHRAE handbook—Fundamentals (I-P)*. Ashrae.

Bradshaw, V. (1993). *Building control systems* (2nd ed.). Wiley.

Grondzik, W. T., & Kwok, A. G. (2019). *Mechanical and electrical equipment for buildings* (13th ed.). Wiley.

Lindsey, R. (2009, January 14). *Climate and Earth's Energy Budget*. Nasa.gov; NASA Earth Observatory. https://earthobservatory.nasa.gov/features/EnergyBalance

Image Credits

Fig. 6.3: Copyright © by Kropsoq (CC BY-SA 3.0) at https://commons.wikimedia.org/w/index.php?curid=692415.

Fig. 6.8: Copyright © 2023 by Martin Gold (CC BY-SA 3.0) at https://commons.wikimedia.org/wiki/File:EM_spectrum.svg.

Fig. 6.10: Copyright © 2016 Pixabay/Pexels.

Fig. 6.13: Vaughn Bradshaw, Building Control Systems, pp. 26. Copyright © 1993 by John Wiley & Sons, Inc.

Module 7

Thermal Properties of Building Materials

Martin Gold

You say to brick, "What do you want, brick?" Brick says to you, "I like an arch." If you say to brick, "Arches are expensive and I can use a concrete lintel over an opening. What do you think of that, brick?" Brick says: "I like an arch."

—Louis Kahn—circa 1960s

Introduction

Much as Louis Kahn's brick prefers an arch—to always be in compression—materials also have thermal properties (their responses to heat), that can be much different across typical building materials. Heat is almost always moving through the building materials that surround us. As heat naturally seeks to move from an area of high concentration to an area of low concentration (hot to cold), heat can be flowing to the inside or the outside of a building depending on the seasonal temperatures, the amount of moisture present, the occupation of the building, and the amount of solar radiation present.

The mechanisms of heat flow, convection, conduction, and radiation, and moist air have been introduced in previous modules. The reader may consider reviewing those concepts as they apply to this section. Here, we will focus on the interaction of these mechanisms of heat flow and how they affect building materials specifically. And, how that heat flow can be controlled or manipulated through the appropriate selection of building materials. It may be obvious that we want to limit the flow of heat to the extent possible, but there are circumstances when we might want to facilitate that heat flow and we will discuss those cases in this module. This module should provide students with an awareness of how heat flow in buildings is calculated and the understanding of how heat flow interacts with building materials. This should provide students with the ability to make informed choices about which building materials might be optimal for a given situation, and gain an intuitive sense of the magnitude of impact various materials will have on building energy use.

Heat Flow

Heat flow can be defined as ***the rate of heat flowing through a building material over a given period of time***. In the United States, the rate of heat flow is measured as British thermal units (Btu) per hour (hr)—Btu/hr. In the case of Figure 7.1, a simple 1 square foot of homogeneous building material is being represented for simplicity. Historically, buildings were typically constructed of homogenous materials—stone, adobe, logs, furs, snow, etc. Modern constructions, however, are usually much more complex with layers of materials that form a systemic barrier between inside and outside. Each layer in a wall, roof, or even window system is generally optimized for its specific function such as durable metal or cementitious siding, structural wood framing or concrete masonry units (CMUs), polyethylene vapor barriers, fibrous or foam insulation, and interior finish materials that may be practical or decorative.

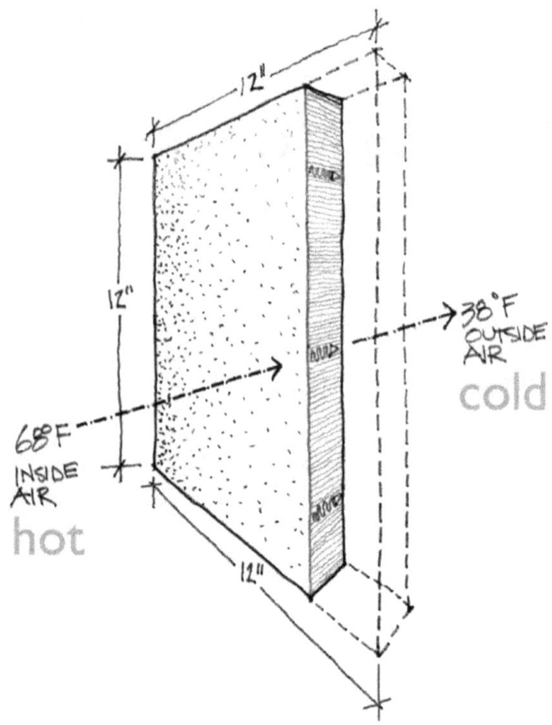

Figure 7.1 Basic Heat Flow Diagram Showing Heat Moving From Warm 68°F to Cold 38°F Through a 1 ft^2 Section of a Nonhomogeneous Material—8-in. Thick Concrete Masonry Unit (CMU).

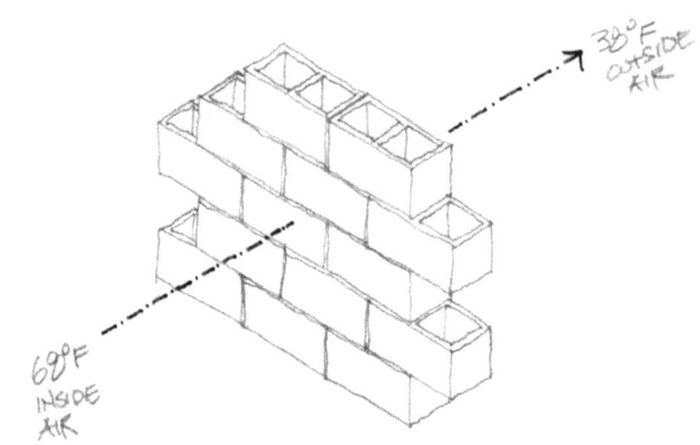

Figure 7.2 Diagram of Heat Flowing Through a Nonhomogeneous Material—8-in. Thick Concrete Masonry Unit (CMU).

Building materials often have different compositional properties with some materials being homogeneous and some with hollow spaces or air cavities such as a concrete masonry units (CMU), brick, or double-wall polycarbonate. These compositional characteristics govern how heat flows through the material and the calculations of heat flow must account for any discontinuities, if present, in the material structure. Therefore, there are two common measures of the capacity of a building material to transmit heat: conductance (C) and conductivity (k).

Measures of Conductive Heat Flow

Conductivity (k) is a measure of the heat flow through a ***homogeneous*** material per inch of thickness. The units of heat flow are Btu/(hr·ft^2·°F). The "k" values are for 1-in. thickness of the material. For example, if you have a homogeneous material such as solid concrete, that is 1-in. thick, then the k value list in a table is the conductivity for that 1-in. thickness of solid concrete. Perhaps you have solid concrete that is 3-in. thick. In that case, one needs to multiply the given k value by 3 (based on the 3-in. thickness) in order to get the correct conductivity for a 3-in. thickness of concrete. Figure 7.1 illustrates a homogeneous material that would have a conductivity value (k).

Conductance (C) is a measure of the heat flow through a ***nonhomogeneous*** material and is given for the thickness listed. Conductance units: Btu/(hr·ft^2·°F). Conductance is preferred for materials such as CMUs that have both solid and voids in the material structure. This also applies to insulating materials as they have air voids that do not perform linearly with regard to heat flow as they get thicker or thinner. Therefore, these materials have conductance ratings rather than conductivity ratings.

R-value (R) measures a particular material's resistance to heat flow for a specific thickness of a material. Resistance to heat flow is the inverse of conduction to heat flow in units: (°F ft^2 hr/Btu). When we want to calculate how much heat might be flowing through a material, we use the conductance. However, when we are designing buildings, and want to know how much a material insulates (resists heat flow), we use the R-value rating. The higher the R-value rating number, the lower the heat flow through the material. The higher the R-value the the better the material is as an insulator.

Yes, but how does insulation work? Heat travels by moving from one molecule to the next and transferring heat as kinetic energy as they bump into each other. The denser the material, the more molecules there are to bump into each other and transmit heat. Conversely, less dense materials have fewer molecules to bump into each other and are therefore poor at transmitting heat. The best insulator is a space without even any air also called a vacuum. High-quality water bottles and thermoses have spaces between the inside and outside "shells" that have most of the air taken out for optimal insulation. After a vacuum (no air), air is the next best insulator.

But insulation used in buildings is a material with more density than air. Why not just leave air in a building cavity? Air spaces in building cavities transmit heat from one side to the other by two mechanisms. (1) radiant heat flow across the air space; and (2) by convective loops occurring in the air contained in the air space. Convection is a naturally occurring phenomenon that causes hot air (less dense) to rise and cold air (more dense) to fall. The actual physics of this phenomenon is that the denser cold air is pulled to the earth by gravity pushing the lighter warm air out and up above it. Typically, air spaces have surfaces with a warm side and cold side which may exchange from summer to winter. The effect of the two different surface temperatures on air in the cavity is that air falls on the cold side and subsequently rises on the warm side. Because the air is trapped in the wall cavity, it forms a "convective loop" of air movement that conveys heat from the warm side to the cold side. In order to stop the convective loop, material, such as glass fiber or polystyrene is placed in the wall cavity. Insulating materials are typically very porous and are effectively mostly air. The porous material holds the air in place so convective loops cannot form. Additionally, the material prevents radiant heat transmission from transmitting across the air space between the surfaces. This trapping of the air in a porous matrix is the mechanism that makes these insulating materials so effective. A 3-1/2-in. air space has an R-value just less than 1.0 while filling that air space with 3-1/2-in. glass fiber batting achieves an R-value of 11 and it can be as high as 15. *Insulation used in buildings can be over 10 times or more effective than an air spaces alone.*

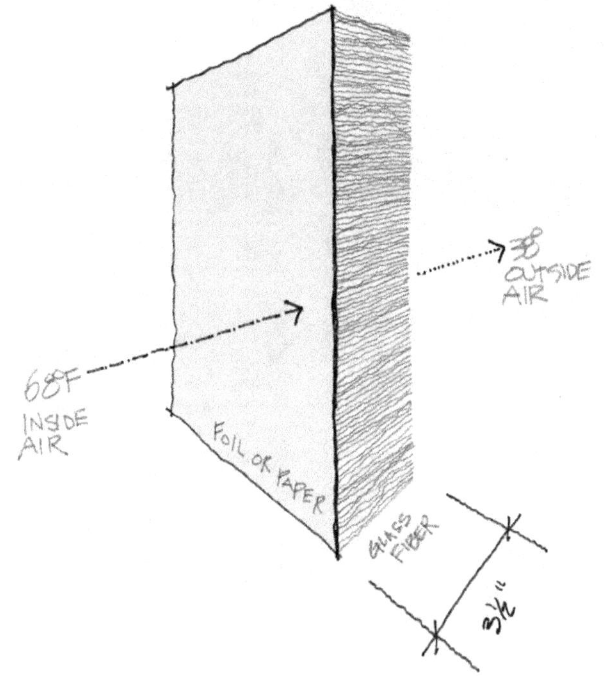

Figure 7.3 Diagram of Heat Flowing Through a 1 ft^2 Homogeneous and High Thermally Resistant Material – 3-1/2-in. Thick Glass Fiber Material.

Architects usually seek to limit the heat flow between the inside and outside to save energy. The single number R-value rating makes it easy to compare the thermal resistance of different materials whether homogenous or nonhomogeneous. One could specify 3-1/2-in. thick glass fiber batting insulation to achieve an R-value of approximately 11. Alternatively, one might choose 2-in. thick high-density polystyrene for an R-value of approximately 10. Or, using a 3-in. thick high-density polystyrene board that could achieve an R-value of just over 15 which is about 1/3 more thermal resistance than the first two options. Depending on the wall cavity thickness, locally available materials, magnitude of the local climate, and material cost, one might choose one material over the other depending on the circumstances.

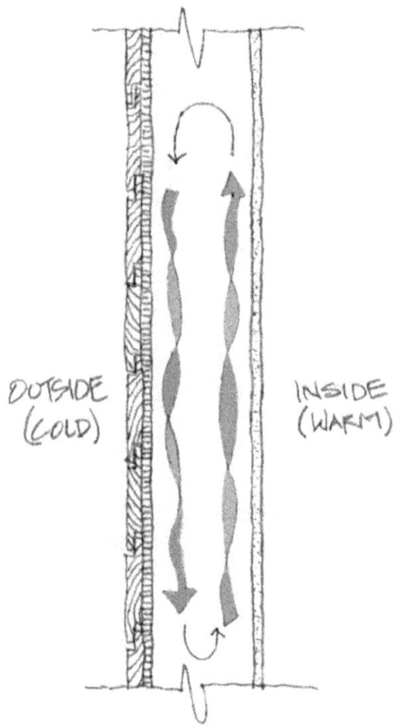

Figure 7.4a Diagram of Typical Wall Cavity with Convective Loop (Left).

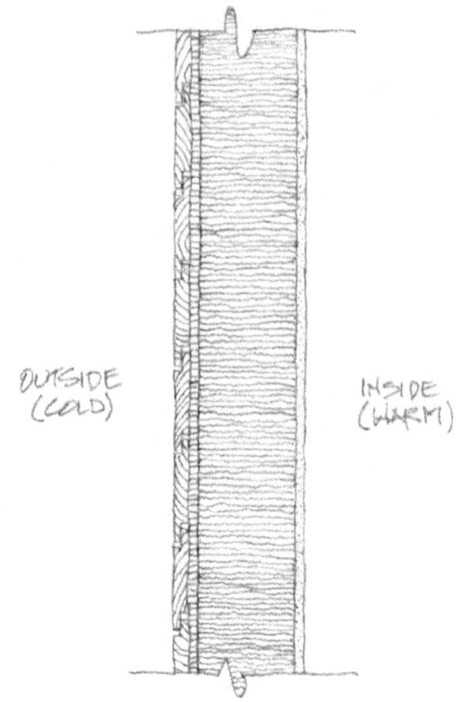

Figure 7.4b Diagram of Typical Wall Cavity with Insulated Wall (Right).

Material	**Thickness**	**R-value**	**Comment**
Glass fiber	3-1/2^2	11	Commonly used
High-density Polystyrene	2^2	10	Compact alternative
High-density Polystyrene	3^2	15	Better for harsh climates
Air space	3-1/2”	1	Why we need insulation
Single-pane glass	1/8^2	1	Poor as an insulator

So far, we have been considering the thermal properties of individual building materials. When materials are assembled, they are thermally more complex than the simple sum of the individual parts. For example, how air forms a convective loop that can wick heat away in a wall cavity while otherwise being considered a good insulator. When we consider the thermal properties of walls as composite assemblies we use the metric U-value.

U-value (U) is a measure of how much heat flows through a wall assembly that is made up of multiple components with units: Btu/(hr·ft^2·°F). The U-value is the inverse of the sum of the R-values and includes the impact of air spaces and radiant heat flow within a wall assembly. U-value is an important metric as it is used to calculate how much heat is flowing through a given wall assembly. If we want to model a building's thermal properties using a computer model, or calculate the amount of heat a building might exchange with its environment, we use the U-value of the wall assembly as the primary thermal metric for composite wall systems.

The U-Value Tells Us How Much Heat Can Be Flowing Through 1 ft2 of a Building Material in 1 hr.

Composite Wall Assemblies

Walls in modern construction are almost always made up of a layered assembly of materials each optimized for factors that a wall must achieve for utility. For example, wood or steel framing has great structural properties and gypsum board is a great interior surface but poor on the exterior. Wood or metal siding can be great as a weather barrier but typically it is thin and not viable structurally. Unless one is renovating a historic masonry building, building a log cabin, or constructing an igloo, they will be using a composite wall assembly. Architects should consider all the opportunities for what those components of the wall, roof, door, or window might be considering design, cost-benefit, and the thermal properties toward effectiveness and sustainability.

Air Films at Building Surfaces. On the inside and outside surface of each wall assembly, an air film will form as molecules are held by the pours of the material at the molecular scale even for very smooth surfaces such as glass. Air films provide a thin insulating layer at the material's surface. For interior air surfaces, the air is typically still so the air film is largely maintained. On the exterior of a building, wind or breezes can carry this insulating layer away. In cold windy climates, this can have a substantial impact on drawing heat out of a building assembly.

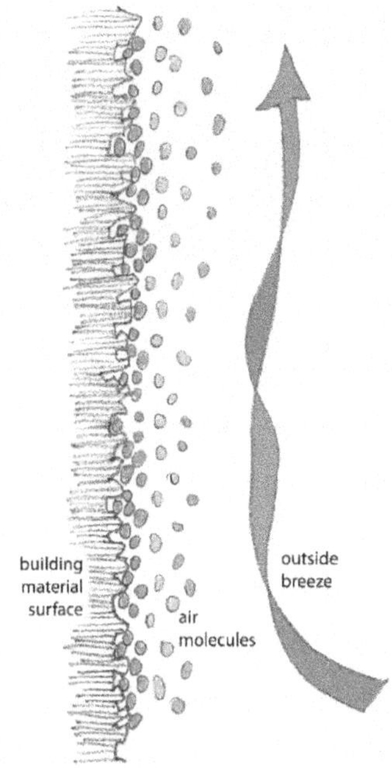

Figure 7.5 Diagram at Molecular Scale of a Wall Surface Illustrating Breezes Removing Heat from a Building Surface.

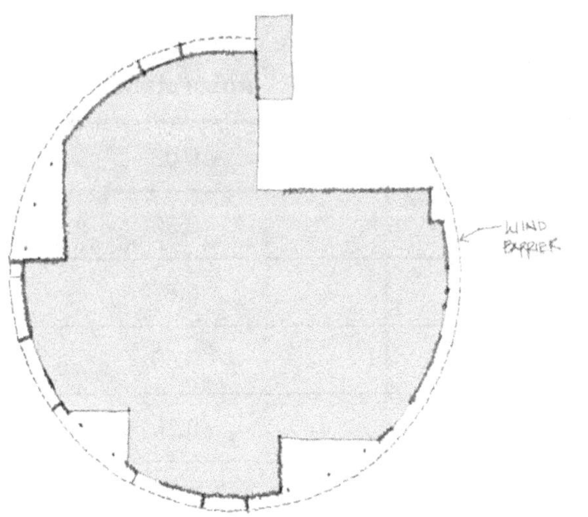

Figure 7.6a Public Library in Viikki, Finland by ARK-House Architects Utilizing a Double-Skin Wind Break to Prevent Wind-Driven Heat Loss on Building Surfaces—Plan (Left).

Figure 7.6b Public Library in Viikki, Finland by ARK-House Architects Utilizing a Double-Skin Wind Break to Prevent Wind-Driven Heat Loss on Building Surfaces—Photo (Right).

The Viikki Library utilized a glass wind-break around the building that is also inhabited space in particular locations. This is an excellent example of technology deployed as design by understanding basic thermal principles, and the value of a separated building skin as an energy savings strategy while carrying that further as a spatial design element to expand the pallet of experiences in the library by providing intermediate inside/outside useful spaces. Note from the image that operable windows are used to passively control the intermediate space using natural ventilation (open) or closed to leverage the greenhouse or sunspace effect to achieve a warm space on cold days without mechanically heating them.

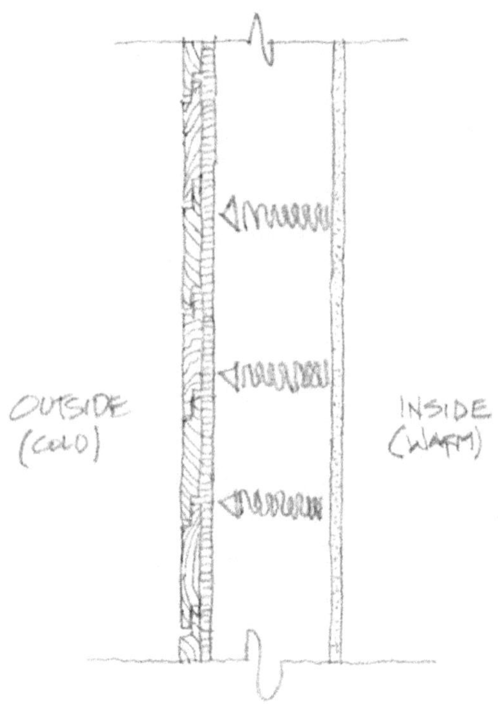

Figure 7.7 Diagram of Radiant Heat Flow Within a Wall Cavity.

Radiant Heat Flow in Building Air Spaces

Radiant heat transfer occurs on the exterior and interior surfaces of buildings, wall cavities, and attics as the material on the warm side of the space "radiates" heat to the cold side of the space. The mechanism of heat transfer is electromagnetic radiation which is the same way heat arrives on earth from the sun. So long as there is a line of sight between the source and receiver, electromagnetic heat can transfer across space. How much heat is radiated over a given period of time depends on the emissivity of the source surface and the absorptance of the receiving surface.

Emissivity (ε) is a ratio from 0 to 1 that indicates how well a material can transmit electromagnetic energy. As a material's emissivity approaches 1 it is very effective at radiating heat. As the emissivity approaches 0 the material does not radiate heat well.

Absorptance (A) is a ratio from 0 to 1 that indicates how well a material can absorb electromagnetic energy. As a materials absorptance approaches 1 it is very effective at absorbing heat. As the absorptance approaches 0 the material is relatively poor at receiving heat.

Table 7.1 Emissivity and Absorptance of Common Building Materials

Material	Emissivity	Absorptance
Aluminum	0.03	0.09
Asphalt	0.88	0.90
Brick	0.90	0.65
Concrete	0.91	0.60
Glass	0.95	0.10
Paint (white)	0.90	0.16

Table 7.1 shows the emissivity and absorptance of common building materials. If we are seeking to limit heat flow through a wall or roof system, we would want to choose materials that restrict heat flow from all the mechanisms: convective, conductive, and radiant. In the case of radiant heat flow, we would want materials that are poor emitters and poor absorbers of radiant heat. Looking through the table (Figure 7.8) there is one material that is superior to the others in both emissivity and absorptance—aluminum. Often one will see building materials with aluminum backing (the side that would face an air space). This shiny side limits the radiant heat flow that can occur across an air space. Attic spaces will often have aluminum foil-backed insulation to reduce radiant heat exchange in addition to thermal resistance. According

to the US Department of Energy, in hot climates, radiant barriers in attics, at the underside of the roof, can reduce cooling costs by 5% to 10%.

If we live in a hot climate, we might want to paint our dwelling white. As can be seen from the table, white paint is poor at absorbing radiant heat while it is also quite good at emitting radiant heat. If we are trying to shed heat to the environment around us, white is a good choice. Consider that a white T-shirt is a good choice on a hot day for the same reasons.

Heat Capacity of Building Materials

Building materials will hold heat for a period of time depending on their specific heat and their density. Generally, the higher the density of a material, the higher the heat capacity and ability to hold heat. From Table 7.2, we can see that water has the highest thermal capacity of the materials listed with steel, concrete, and brick having a diminishing capacity to hold heat. Wood is quite low relative to the others and tends to shed heat relatively quickly.

Table 7.2 Heat capacity for Common Building Materials, From High to Low

Material	Specific heat × density (Btu/°F · ft^3)
Water	62,587
Steel	56,005
Concrete	29,875
Brick	20,246
Wood	11,652

This physical phenomenon of thermal capacity is important in terms of energy-efficient design and appropriate responses to climate. For example, in hot humid climates, the daytime and nighttime temperatures are similar, perhaps only ranging from 5°F to 10°F from day to night during summer months. In this case, building materials that release their heat to the environment faster such as wood, or thin metals would be preferred.

In climates with high diurnal temperature swings ranging from 40°F to 60°F, such as hot arid climates, materials with a high heat capacity perform better as they hold the heat from the day and lose that heat slowly at night. Using thermal mass to stabilize interior temperatures in these climates provides two important phenomena to moderate the high diurnal outdoor temperature swings—(1) thermal time lag and (2) thermal moderating effect.

Thermal time lag occurs as it takes time for a material with high thermal capacity to heat up and then cool down. The result is that the highest temperature throughout the material comes much later in the day than the high exterior temperature.

Thermal moderating effect occurs primarily as a result of the thermal time lag noted above. Because high thermal mass materials are very slow to transmit external temperature changes to the interior, the magnitude of the outside temperatures is moderated by the material.

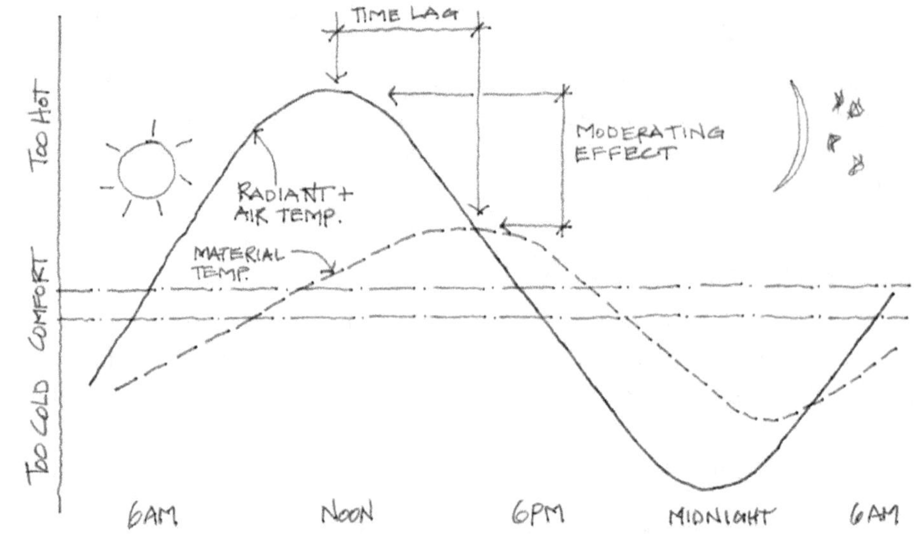

Figure 7.8 Thermal Time Lag and Moderating Effect Achieved with Building Materials That Have a High Heat Capacity.

Estimating the U-Value of a Wall Assembly

When considering wall systems for a building, architects should have a working knowledge of how to specify assemblies that are thermally effective and suited to the climates in which they will inhabit. Durability, availability of materials, design aesthetic, and of course thermal performance should be considered. This section walks us through calculating the U-value for a wall assembly and then checking the magnitude of improvement by using insulation or leaving an open-air space in the wall cavity.

For this example, we will need to find R-values of various building materials. This information is best searched on the internet as new materials and updated data are published regularly. The information can be found on manufacturer's websites or compiled in summary tables. The sources below are good options for gathering the thermal properties of particular building materials or one can just search for the R-value of a particular material and usually find relevant values.

- *Mechanical and electrical equipment for buildings* (MEEB), 12th ed. Free download.
  - Click on the link below, and scroll to "DOWNLOAD OPTIONS," then select your preferred format.
  - https://archive.org/details/Wiley.Mechanical.12th.Edition.Oct.2014.ISBN.1118615905
  - In the download, scroll to the tables in Appendix E (E.1, E.2, E.3, E.4) pages 1649–1671.
- Archtoolbox—R-values of typical building materials
  - https://www.archtoolbox.com/r-values/
- R-values of typical building materials
  - https://www.ahfc.us/iceimages/manuals/building_manual_ap_1.pdf

Recap: The U-value (U) of a wall assembly is the inverse of the sum of the R-values (R_1 R_2 R_3...).

$$U = 1 \div \sum (R_1\ R_2\ R_3 \ldots)$$

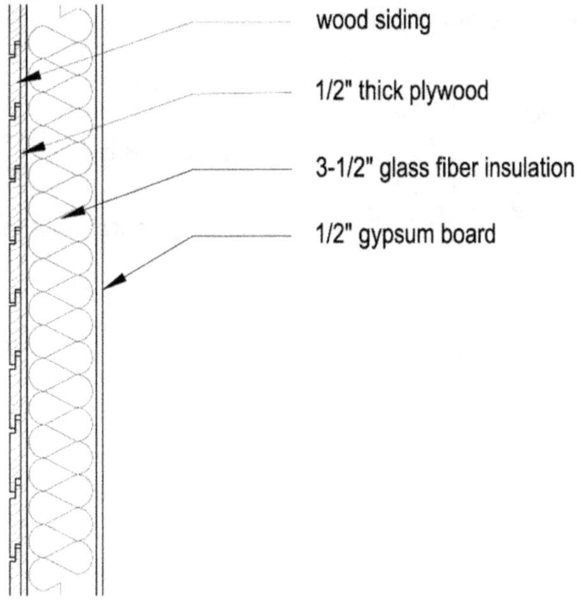

Figure 7.9 Typical Wall Assembly; Wood Siding Over Wood Studs with Glass Fiber Insulation and ½" Gypsum Board (Typical of 1960-Present).

STEP 1: Estimate the R-value of the air film on the exterior of the wall assembly.

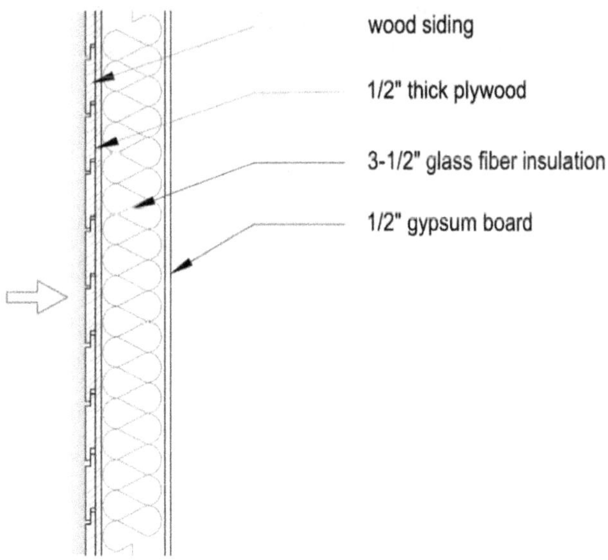

Figure 7.10 Outside Air Film of Wall Assembly r-value = 0.17.

This value can be found in the links provided. In the MEEB download, find the value on page 1668. In the case of an outside air film, use the "moving air" value for "winter wind."

STEP 2: Estimate the R-value of the wood siding.

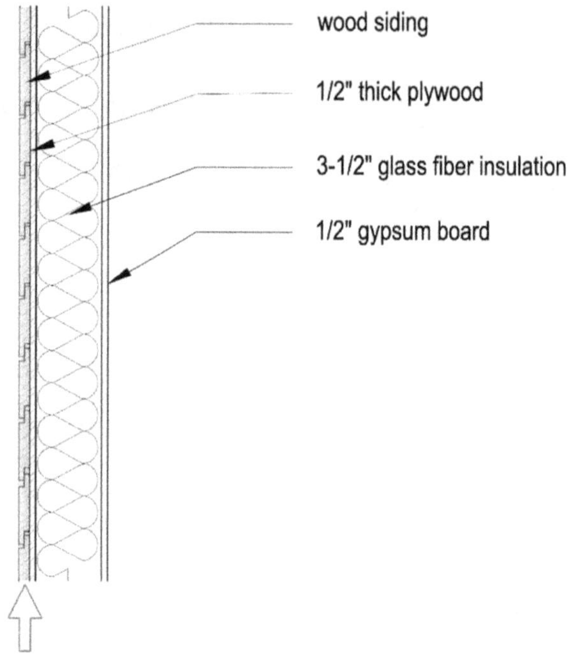

Figure 7.11 Wood Siding r-value = 0.81.

One could search for this value on the internet or use the links above. There may be slight differences in values depending on the density of the wood so expect some variation.

STEP 3: Estimate the R-value of the plywood substrate (sheathing).

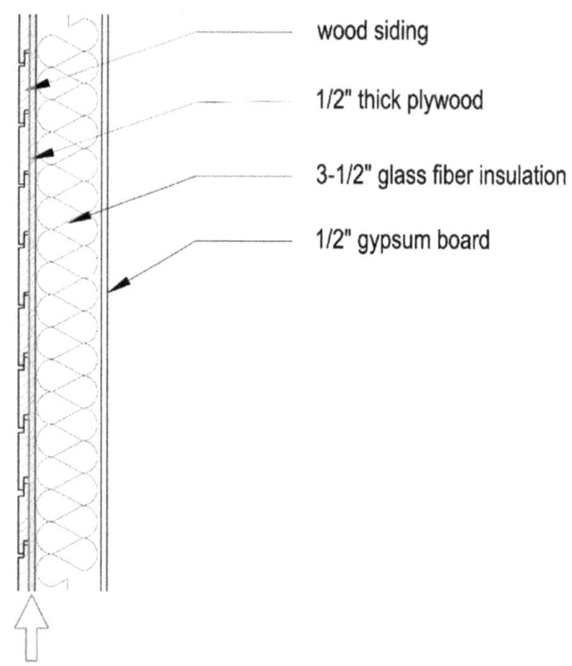

Figure 7.12 Plywood Substrate or Sheathing r-value = 0.62.

Again, this information can be found from many places and could vary slightly between sources and depending on the exact density of the plywood material. Without specific information, make an educated guess at which value to use.

STEP 4: Estimate the R-value of the glass fiber insulation.

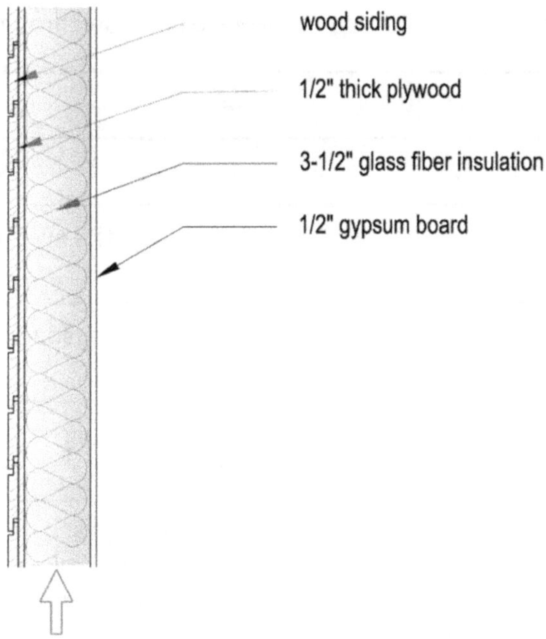

Figure 7.13 3-1/2-in. Thick Glass Fiber Batt Insulation r-value = 15.

Here again, there are different densities of glass fiber associated with their resulting R-values. R-11 is considered the minimum insulation value for a wall cavity and it could be higher in some states with extreme winters. The author used 3-1/2-in. "mineral fiber" with a density of 1.2-1.6 lb/ft^3 which yielded an R-value of 15 (from MEEB Table E.1 on page 1651).

STEP 4 (continued): Estimate the R-value of the wall without insulation—air space only.

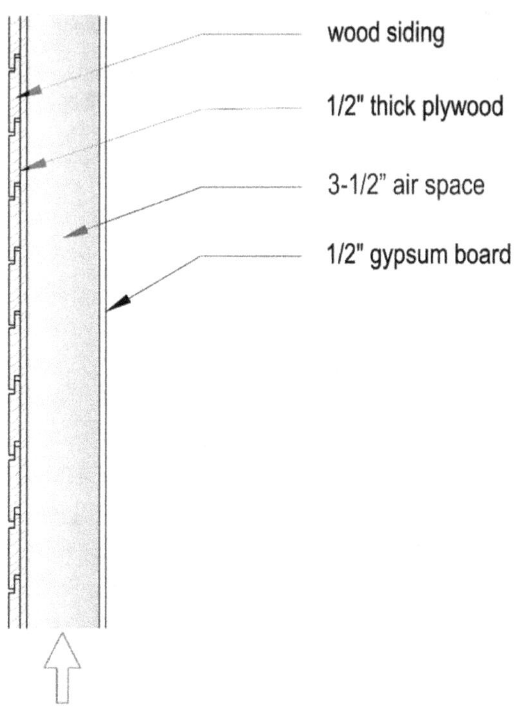

Figure 7.14 3-1/2-in. Air Space r-value = 0.91.

In order to calculate the air space, some information is needed to describe aspects of the air space such as the direction of heat flow, nominal temperature differences, and the emittance of the materials enclosing the air space. Referencing MEEB page 1670, one can follow through the table:

Position of air space	Vertical
Direction of heat flow	Horizontal
Mean temperature*	50*—this is the average of high and low for a given day (winter)
Temperature difference	30*—across airspace 65°F inside and 35°F outside. (estimates)
Emittance	0.82—assuming no foil or radiant barrier is applied
Result	**R-value = 0.91**
**The values noted are the most commonly used but one can interpolate for more refined data.*	

It should be noted that the ***R-15 insulation provides 15-times the thermal resistance to heat flow compared to the air space alone*** which is approximately R-1. Also, one can see that small magnitudes of increased insulation, R-value 0.5 or less, have little impact on the overall insulating value once a high R-value insulating material is added.

STEP 5: Estimate the R-value of the interior wall material—1/2-in. gypsum wall board.

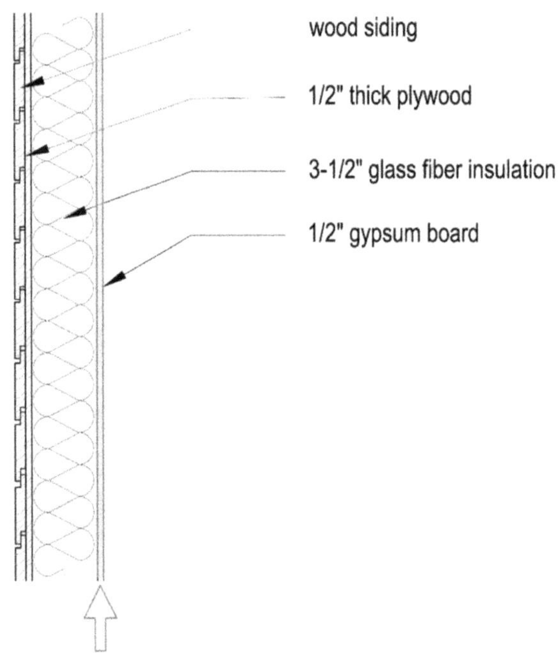

Figure 7.15 1/2-in. Gypsum Wall Board r-value = 0.45.

R-values for the gypsum board are readily available through web searches or from the MEEB download.

STEP 6: Estimate the R-value of the air film on the interior of the wall assembly.

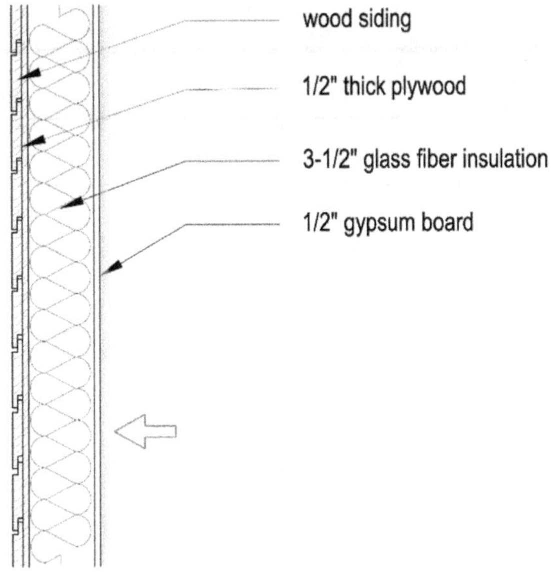

Figure 7.16 Inside Air Film r-value = 0.6.

This value can be found in both the links provided. In the MEEB download, find the value on page 1668. In this case, the orientation of the surface is important (vertical for a wall) and the direction of heat flow (horizontal for a wall). If this was a roof system, those parameters would be different.

STEP 7: Sum the R-values and then take the mathematical inverse to determine the U-value of the wall assembly.

Table 7.3 Summary of R-Values

Material	R-value	Notes
Outside air film	0.17	
Wood siding	0.81	
½² plywood	0.62	
3 ½² insulation	15.0	0.91 airspace
½² gypsum board	0.45	
Inside air film	0.65	
∑ of R-values	17.7	With 3.5 air space.

Table 7.4 Conversion From the Sum of the R-Values Into the Composite Assembly U-Value.

Assembly U-value
$U = 1 / \sum$ R-values
$U = 1 / 17.7$
U-value = 0.0565

From Tables 7.3 and 7.4, the math is straightforward in terms of calculating the U-value for the wall assembly once the various R-values have been determined.

Heat Flow Through Building Materials

Now that we know what the U-value is for our wall assembly that will be used to enclose a building, we want to know how much heat will transmit through that assembly for a given indoor and outdoor temperature. For example, if we establish design temperatures of say 30°F outside and 68° inside, we could then calculate how much heat is lost on an hourly basis and then size mechanical equipment and estimate how much it might cost to heat the building for a season. The formula shown in Figure 7.22 below is used to calculate heat loss in buildings.

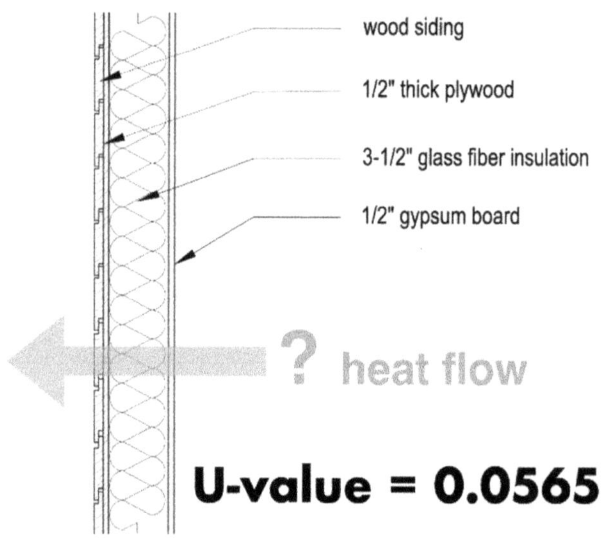

Figure 7.17 Concept of Heat Flow From Inside to Outside in a Winter Condition.

Table 7.5 Formula for Calculating Heat Flow Through a Building Assembly With a Given U-Value

Q = Btu/hr
U = Thermal transmittance (Btu/hr·ft^2·°F)
A = Surface area in contact (ft^2)
Δt = Temperature difference between objects/surfaces (°F)
Q = U Btu/hr·ft^2·°F **× A**ft^2 ×**Δt**°F

Having established a design temperature of 30°F outside and 68° inside, the difference between those temperatures is the delta-t (D*t*), in this case 38°F. We have calculated the U-value (0.0565) and that leaves the area of the wall assembly (A in square feet) as the only unknown variable. For now, we will use 1 ft^2 (1 sf) as the unit area. Then we can simply multiply the resulting heat loss in (Q in Btu/hr) by the total area of the wall assembly and find the total heat loss.

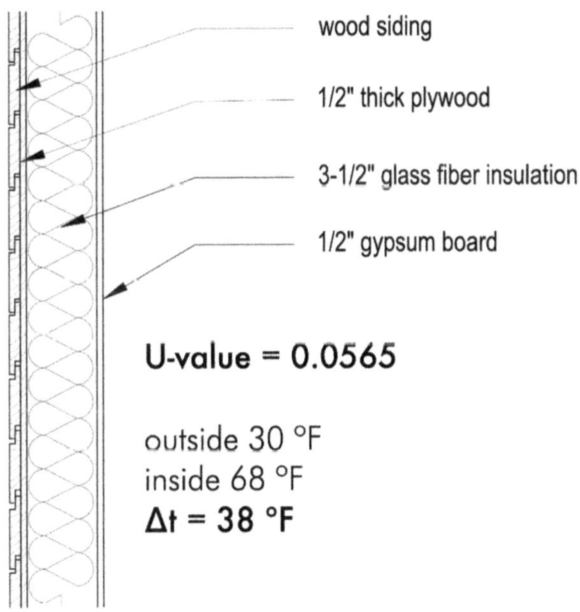

Figure 7.18 Heat Flow Through Wall Assembly Summary With Exterior and Interior Temperatures and the U-Value of the Wall Assembly.

Table 7.6 Calculation of Heat Loss in British Thermal Units (BTUs) per Hour

Q = 0.0565 Btu/hr·ft^2·°F **× 1.0**ft^2**× 38**°F
Q = 2.147 Btu/hr for each square foot of wall area

Comparing the insulated wall to that of a typical window that might be placed in the wall, we can see there is a substantial difference in heat flow between the two building assemblies—window and wall. The window, with an R-value typically less than 1.0, allows 11 times more heat to flow through it than the insulated wall. In researching window R-values, one will find that even with double-pane insulating glass, the R-values rarely exceed 2.0.

Table 7.7 Calculation of U-Value for a Simple Single Pane Window

Material	**R-value**
Outside air film	0.17
¼" glass	0.90
Inside air film	0.65
∑ of R-values	1.72
Q = 0.581 Btu/hr·ft^2·°F **× 1.0**ft^2**× 38**°F	
Q = 22.01 Btu/hr for each sf of glass area	

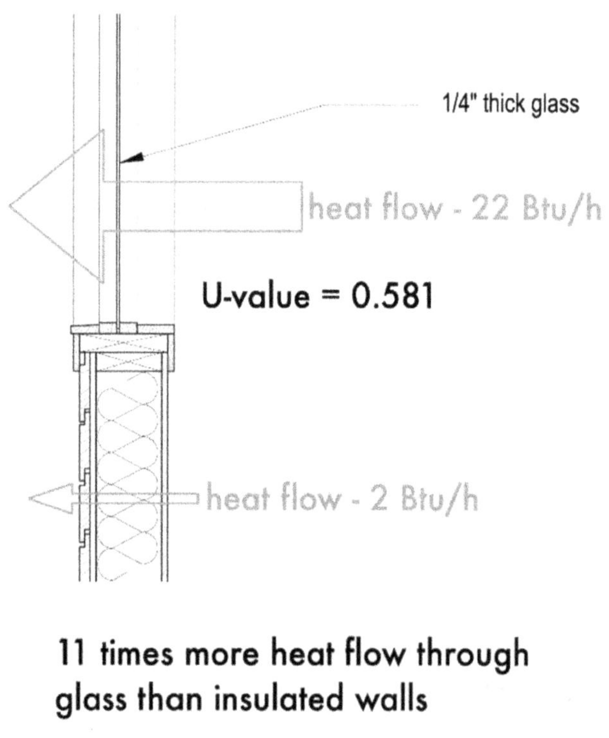

Figure 7.19 Summary of Heat Flow Through a Typical Wall and Window Assembly for a Given Set of Temperatures.

Moisture and Thermal Properties of Building Materials

Moisture is present, to some degree, in the building materials used in construction and can vary depending on the climate, porosity of a material, and the temperature of the material relative to the surrounding air. Moisture in the air is discussed in the module on psychrometry. Here, we will discuss the effects of moisture on heat transfer in building materials and assemblies.

When materials are damp, wet, or saturated with moisture, they become denser as water molecules fill voids in the porous material. Subsequently, heat flows more easily through the materials. When the relative humidity is high, it can reduce the effectiveness of porous insulating materials. Figure 7.27 shows how a few representative insulating materials lose resistance as relative humidity is increased. At 40% relative humidity, materials typically yield 100% of their rated R-values. As humidity increases, thermal resistance can be diminished to well over half of the dry performance depending on the material.

Table 7.8 Relationship of Relative Humidity and Reduction in Thermal Resistance of Insulating Materials

Material	**Relative humidity (Rh)**			
	40%	**60%**	**80%**	**100%**
Glass fiber	100%	93%	83%	48%
Mineral fiber	100%	100%	95%	90%
Expanded Polystyrene (EPS)	100%	97%	94%	85%
Extruded Polystyrene (XPS)	100%	97%	97%	91%
Phenolic foam	100%	88%	52%	35%

Condensation was introduced in *Module 5 Thermal Properties of Air and Psychrometrics* as a phenomenon where liquid water can form on surfaces by "condensing" out of the air. This is also true for solid water condensing directly as ice if the surfaces are cold enough. For this to happen, there must be moisture in the air and the air must be warmer than the surfaces. The moisture can come from inside a building from people directly (respiration and perspiration), appliances such as dishwashers or washing machines, and from cooking or bathing. In hot humid climates with daily relative humidity regularly reaching near 100%, the moisture can come primarily from outside. In cold or arid climates, the moisture typically comes from within the building most of the year.

Visible condensation forms on building surfaces such as walls, windows, exposed pipes, floors, or duct vents and typically we can see this. This moisture can degrade materials and cause mold to form on material surfaces. When this occurs on floors, it can cause slip hazards and can be dangerous. Thermally, it can wick heat away as the water layer is a much better conductor of heat than a dry air film.

Concealed condensation arises from water vapor's natural tendency to move from an area of high vapor pressure to an area of low vapor pressure. This pressure gradient can transmit through most porous building materials and through cracks and gaps in construction. Once the vapor is in a wall cavity, if one of the surfaces is cold (below the dewpoint temperature) water can condense on the surfaces in the wall cavity as a liquid. The wet materials will be much more conductive to heat. If fiberglass batt insulation gets wet, it can collapse in the wall cavity slumping to the bottom and leaving air space gaps with much less thermal resistance.

Vapor barriers are an effective strategy to reduce moisture in wall cavities. Typically, the vapor barrier goes on the warm side of the wall, usually just under the finish material. In cold climates, the warm side is the inside for most of the year. In hot and especially humid climates, the warm side would be the outside. As one can expect, the warm side can switch for periods of time seasonally. Also, the wall assembly needs to have some ability to shed moisture from seasonal changes, potential water leaks, or condensation in the wall cavity leaving one to choose which side to put a vapor barrier.

Vapor barriers are measured by their permeability in "perms"—1 grain of water vapor per hour, per square foot, per inch of mercury. The lower the "perm" rating, the higher the resistance to vapor. Building codes in the Northeast US call for a class II vapor barrier system of not greater than 1.0 perms or better. Vaper barriers are either class I (< 0.1p) or class II (0.1p to 1.0p) or class III (1.0p to 10p). Class II and class III are considered semipermeable. A "perm" rating above 10p is considered a vapor permeable membrane.

Table 7.9 Vapor Permeability Ratings for Common Building Materials

Material	**Perms**	**Class**
Aluminum foil (0.35 mil)	0.05	I
Polyethylene (2.0 mil)	0.16	II
Polyethylene (2.0 mil)	0.08	I
3 Coats oil paint on wood siding	0.30	II
15 lb asphalt	1.0	II
8^2 Concrete block (unpainted)	2.4	III
¾^2 Plaster on metal lath	15	N/A
½^2 Gypsum wall board (unpainted)	48	N/A
½^2 Gypsum wall board (2 coats paint)	5	III

Controlling moisture can be difficult given the variability of climates, proliferation of air-conditioning, and need for relatively moisture free building assemblies for optimal thermal performance. Proper use of vapor barriers and ample ventilation are two strategies that can limit moisture borne degradation in wall assemblies.

Summary

This module covers important thermal properties of building materials and assemblies leading to their composite thermal performance. Points of knowledge that should be carried forward include an awareness and understanding of the metrics R-value, U-value, emissivity, absorptance, and the heat capacity of building materials. Readers should have an awareness of how those factors influence the amount of heat flowing through a building assembly. It is important to be knowledgeable about what strategies might be used in different climates to leverage materials with specific thermal properties toward providing comfort and limiting reliance on mechanical heating or cooling.

A typical wall can be 15 times or more effective at reducing heat flow with insulation compared to just an air space. Windows have very low R-values and subsequently high U-values resulting in high quantities of heat transfer during hot or cold seasons. Many building codes inherently limit the amount of window area a building can have by requiring high average insulation values. If a designer wants to have more windows, hopefully operable for breezes and ventilation, they will need to maximize the R-values of the opaque walls, ceiling and perhaps insulate the floor or slab to subsequently allow for more window area. The higher the R-values in opaque walls and roof systems can counteract the low R-values of windows resulting in the potential for significantly greater window area. Similarly, an intuitive understanding of water vapor flows should be developed for thermal, indoor air quality, and material maintenance points-of-view.

Developing an intuitive sense of the properties of building materials can be nurtured through visiting buildings, touching the materials (are they warm or cool, wet or dry), noting the material, and researching their thermal properties online. One might even start building their own database of preferred materials and their thermal specifications.

References

Archtoolbox. (2022, February 27). *R-values of Insulation and Other Building Materials – Archtoolbox*. Www.archtoolbox.com. https://www.archtoolbox.com/r-values/

ARK-house architects. https://www.ark-house.com/ark-house-en/ajankohtaista.html

ASHRAE. (2021). *ASHRAE handbook—Fundamentals (I-P)*. Ashrae.

Bradshaw, V. (1993). *Building control systems* (2nd ed.). Wiley.

Department of Energy. (2022). *Insulation*. Energy.gov; US Department of Energy. https://www.energy.gov/energysaver/insulation

Grondzik, W. T., & Kwok, A. G. (2014). *Mechanical and electrical equipment for buildings*. (12th ed.). Wiley https://archive.org/details/Wiley.Mechanical.12th.Edition.Oct.2014.ISBN.1118615905

Lindsey, R. (2009, January 14). *Climate and Earth's Energy Budget*. Nasa.gov; NASA Earth Observatory. https://earthobservatory.nasa.gov/features/EnergyBalance

US Department of Energy. (n.d.). *Radiant Barriers*. Energy.gov. https://www.energy.gov/energysaver/radiant-barriers

Wang, Y., Zhang, S., Wang, D., & Liu, Y. (2023). Experimental study on the influence of temperature and humidity on the thermal conductivity of building insulation materials. *Energy and Built Environment*, *4*(4), 386–398. DOI: 10.1016/j.enbenv.2022.02.008. https://www.sciencedirect.com/science/article/pii/S2666123322000186

Image Credits

Fig. 7.9: Copyright © by Autodesk Inc.

Fig. 7.10: Copyright © by Autodesk Inc.

Fig. 7.11: Copyright © by Autodesk Inc.

Fig. 7.12: Copyright © by Autodesk Inc.

Fig. 7.13: Copyright © by Autodesk Inc.

Fig. 7.14: Copyright © by Autodesk Inc.

Fig. 7.15: Copyright © by Autodesk Inc.

Fig. 7.16: Copyright © by Autodesk Inc.

Fig. 7.17: Copyright © by Autodesk Inc.

Fig. 7.18: Copyright © by Autodesk Inc.

Fig. 7.19: Copyright © by Autodesk Inc.

Module 8

Natural Systems Site Planning

Those who dwell, as scientists or laymen, among the beauties and mysteries of the earth, are never alone or weary of life.

—Rachel Carson

Introduction

In his seminal and often referenced book *Genius Loci: Towards a Phenomenology of Architecture*, Christian Norberg-Schultz examines the spirit of place—the genius loci. In his phenomenological approach, Norberg-Schultz seeks to find the relationships between the physical world, experience, and human consciousness. He introduces us to the notion of dwelling—not a structure we might live in—but rather the process of dwelling, that is a lifelong interaction with the entwined natural and cultural environments in which we live. He proposes that "dwelling" goes beyond mere shelter and suggests that we live in places with distinct and meaningful character. As architects, we might seek to nurture those distinct elements of place to strengthen our connectivity and to harmonize within the facilities and difficulties that a place might offer. Before we consider our architecture, perhaps we should consider the placeness of the context we seek to inhabit. What are the physical and experiential opportunities for dwelling in harmony with a particular place?

Primitive Dwelling

We might learn much from the study of primitive technology in a particular place that emerged prior to modern technology. Indigenous people responded to their climates often in quite sophisticated ways using the landscape forms, materials, and their ingenuity to create livelihoods that formed community, family, art, and productive cultures.

In one of the harshest environments above the Arctic Circle, Inuit cultures of northern Canada and Greenland fashioned dwellings from the snow and sea ice that would form in winter. Traveling out onto the frozen seas, hunting for seals and polar bears, the Inuit constructed sophisticated dome "Igloo" structures from the snow and sea ice—the only materials available. Because snow and ice trap air, these materials can become quite good thermal insulators. As fresh, lighter snow accumulates, the insulting value increases. Outside temperatures can be negative 45°F and can drop to as low as negative 80°F. Depending on the size, number of people, and any heat source in the igloo, the interior temperature can range from 19°F to 61°F.

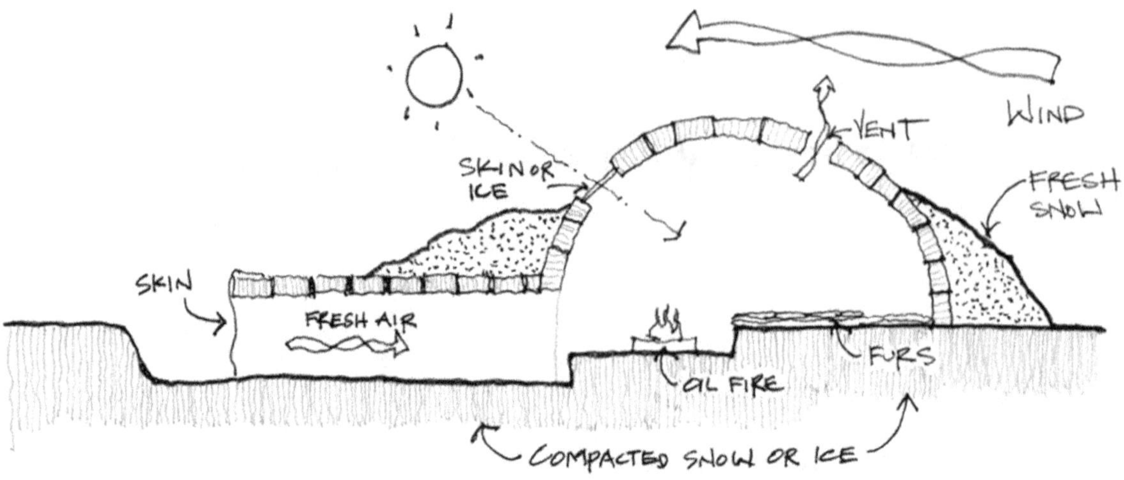

Figure 8.1 Section Diagram of an Inuit Igloo Construction Using Ice and Snow From the Frozen Sea Surface; the Structural Dome Technology, Orientation of Closure to the Prevailing Winds, Digging Into Ground, and Letting Snow Accumulate at the Surface Demonstrates a Sophisticated Knowledge of Structure, Heat Flow, Ventilation, and the Materials.

In northern Europe, near the arctic circle, stone age land dwellers constructed community huts that were connected to gether utilizing logs, sods, and mosses from the forest floor. Unlike the Inuit out on the baron ice sheets these Indigenous people used the forest for protection from the cold north winds in winter and for shade in summer. The structures in Figure 8.2 are from the Kierikki area of northern Finland, just south of the Arctic Circle with annual temperatures ranging from negative 22°F to 86°F. Although it was previously believed these stone age dwellers were nomadic, following the climate, it became clear from ongoing excavations dating back to the 1960s, that these people were permanent residents along the Iijoki River, who developed dwellings suitable for winter and summer. The heavy mosses and large ground area relative to the walls would keep the spaces cool in the summer. During the winter, the sloped walls would hold snow for insulation and again, the large ground area would remain much warmer than the outside air. Linking the huts together allowed for community interaction and collective activities during the harshest of times.

Figure 8.2 Reconstructed Kierikki Stone Age Huts Nested Within a Pine Forest in Northern Finland Approximately 100 Miles South of the Arctic Circle.

Figure 8.3 Interior Spaces of the Kierikki Stone Age Huts Showing the Community Connectivity, Structure, and Spaciousness of the Dwellings.

Moving closer to the equator, the Indigenous tribes of the Miccosukee and Seminole Native Americans developed the chickee in response to the intense South Florida sun, humidity, and frequent thunderstorms. Interestingly, both "chickee" and "igloo" are the native words for house. In response to this climate, the natives used local materials including palm tree trunks, fronds, and grasses. These lightweight materials shed the heat of the day quickly while providing much needed shade and rain protection. There are no walls thus allowing the faint breezes to easily flow through to cool and dry the space to the extent possible. The floor is raised off the ground to keep it dry during the regular heavy thunderstorms that saturate the ground with water. Often, the chickee structures were built in open areas near water bodies to optimize natural convective flows of cooler air coming off of the lakes.

Figure 8.4 Sketch of a Native American Chickee Typical of Those Found in Florida.

Site and Natural Systems

If we consider, as our ancestors have, how to find comfort and livelihood in challenging environments, while using few resources, we too can find opportunities in the natural environment and use them to our advantage. If we are to construct an architecture—a “third environment” as characterized by James Marston Fitch—we should first consider how that environment is maintained through three levels of environmental control: natural systems, passive systems, and active systems. This prioritization of natural, passive, and then lastly active systems could be applied to light, thermal control, air quality, or sound for example. The diagram in Figure 8.5 illustrates this concept showing that we might first find the natural systems to mitigate environmental extremes, then passive systems to further mitigate stresses, then the remaining environmental load would be met with mechanical systems.

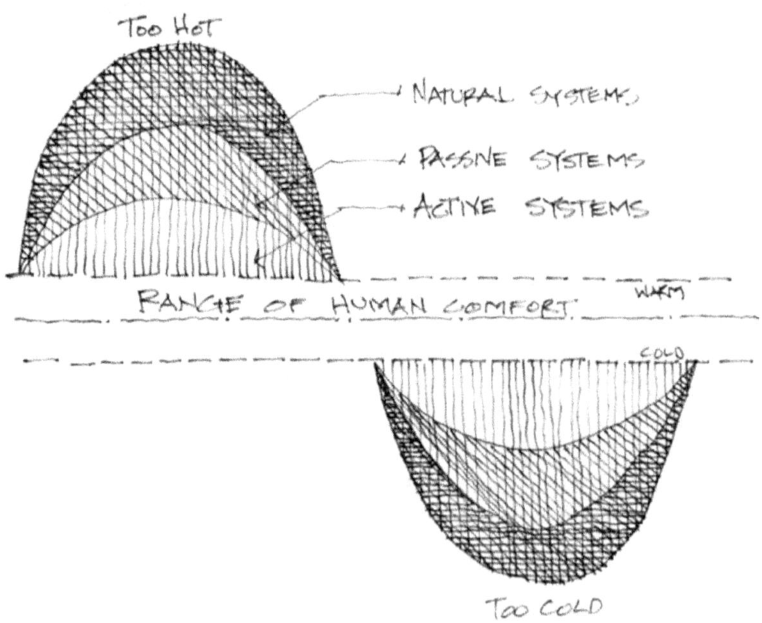

Figure 8.5 Priority for Environmental Control of the Architectural Environment—Thermal Responses.

The opportunities within natural systems rely on our understanding of the environmental elements and how those elements might hold both opportunities or difficulties. For example, Le Corbusier understood the sun can seem evil on a hot summer day with no shade yet can be beautiful and warming on cold sunny day. These facilities and difficulties might be considered generally as climate responses and more specifically at the site scale of design. Issues of consideration:

Climate	Site and natural systems
Sun	Topography
Cloudiness	Geography
Seasonal temperature	Geology
Humidity	Vegetation
Precipitation	Water
Air movement—breezes	Breezes

Climate issues were introduced in module 3 Sun Climate and Weather. Here site issues will be discussed directly as they relate to the general climate while focusing on typical site conditions that may vary with elevation, proximity to water, dense or sparse vegetation, or that might be in more rural or urban environments. In some cases, the local conditions might significantly supersede general climate responses and be counterintuitive. Those cases and general strategies will be presented to the extent possible in this narrative. It is important to use the many climate and local environmental tools available to carefully develop an appropriate understanding of, and strategies for maximizing the natural systems that may be available.

Topography

We use topographic representations of the geography of the places we might inhabit to characterize the natural and constructed artifacts of a place. These representations of course are not the actual place, yet we can take measure of the reality of places through maps, diagrams, and illustrations. These images, documents, and increasingly computer programs such as Google Earth Pro, provide unprecedented information about the recent and past landscape with representations of geography, water bodies, streets, ground level imagery, and history. This is a valuable tool in taking measure of the natural opportunities of a given site.

Figure 8.6 Winter in Ortisei, Italy Nestled in a Valley Between Two Mountains, Remains in Shadow for Most of the Winter; One Can See the Shadow, and How the Snow Has Melted in the Sunnier Areas Yet Still Remains in Shadowed Area of the Village.

Geography

The physical characteristics of place, coastal, mountain, dense urban, open rural, dry, or humid will offer opportunities to reduce environmental loads on our buildings.

In **mountain regions** our access to the sun may be enhanced if we are sited on south facing slope with winter winds blocked by the mountain. Mountain and valley geographies typically have micro-climates that change throughout the day and with the seasons. Where possible, locate on the south rather than north facing slopes (northern hemisphere) and try to avoid long periods of shadow from adjacent mountains.

As the sun rises and warms a mountain face, the air begins to rise causing a prevailing breeze in the valley formation. Around noon, as both mountain faces warm, the rising air will draw breezes up the valley from lower to higher elevations. As the sun transit, the breezes will shift from west-to-east to east-to-west. Finally, after the sun sets, heat will radiate to the night sky, cooling the air, and it will then pool in the valley making the valley the coldest area in the late night and early morning hours.

Natural siting strategies to reduce thermal loads suggest sites partially up the hillside to avoid cold air that will pool in the valley at night. Also, to avoid the higher elevations or the top of the mountain that will experience higher winds. If southern facing hillside is possible, that would be preferred to allow passive solar heating in winter. In high elevations, air becomes less dense and expands which is also a cooling process, at low altitudes the air is denser (compressed) and warmer. This adiabatic process of warming (compression) and cooling (expansion) is complimentary to natural convective flows and can be a factor in the microclimates of mountain contexts and can generate rising and falling gyres in mountain-valley locations.

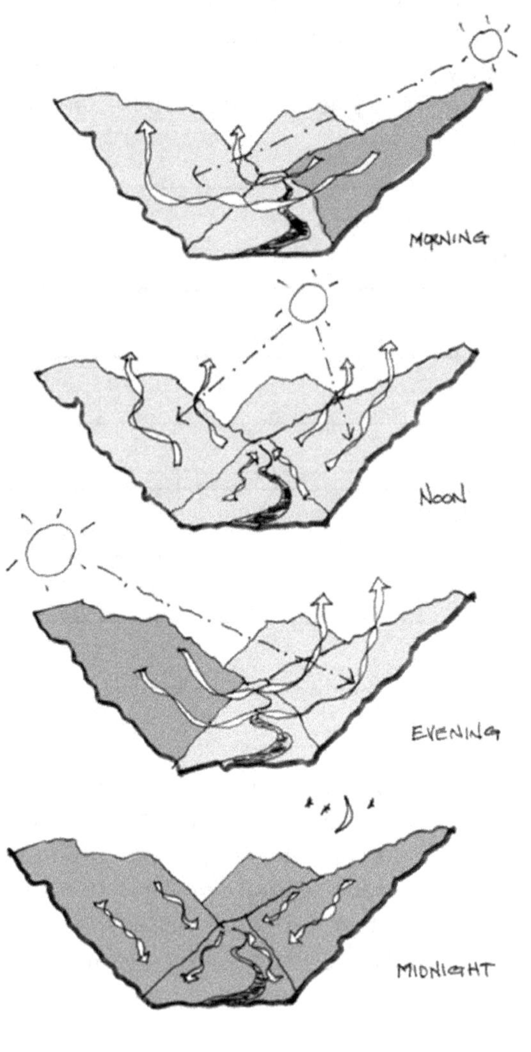

Figure 8.7 Diagram of Air Flow Patterns in Mountain Valley Geography: Morning, Noon, Evening, Night.

Figure 8.8 7123 Thermae, Vals, Switzerland by Peter Zumthor Looking East-Northeast in February.

Peter Zumthor's famous "Thermae" baths in Vals, Switzerland, and associated hotel, is well situated being elevated above the lower valley floor—up and out of the pooling cold night air in the lower valley. The site is on a southeast facing slope taking advantage of the warming low morning sun. The dark granite walls will absorb and hold solar heat. At 37°N latitude, the sun is typically at a relatively low angle in the sky primarily striking and warming the vertical surfaces of the granite walls throughout the day. At night, the sodded roof, with snow cover in winter, insulates the building from losing heat through radiation to the cold night sky. The building is embedded in the mountain deflecting and letting the cooler evening breezes falling down the mountain face pass over the building, while opening to the valley to allow the warmer daytime breezes into the exterior spaces. Known for his outstanding work with materiality and light, Zumthor is also keenly aware of thermal relationships between his architecture and its context.

Alvar Aalto's Experimental House in Muuratsalo on Lake Kivijärvi is an excellent example of embracing natural systems to moderate the cold winters and warm summers of central Finland. Temperatures at the site can average 10°F in winter to 80°F in summer with record lows reaching negative 37°F and highs of 94°F. The protected courtyard faces south with open spaces capturing the summer breezes coming off the adjacent lake. The home is nested into a natural depression in the mountain that protects the structure from the cold winter winds that come from the north. At this latitude (62° NL), the site is only 4° south of the arctic circle and the sun is only in the sky a couple of hours a day in the winter. In summer however, the sun will strike the building for up to 20 hours a day. Nesting the dwelling in the woods provides needed shade during the long and sometimes hot summer days while also diffusing strong winter winds that can blow consistently at an average speed of 9 miles per hour in open spaces.

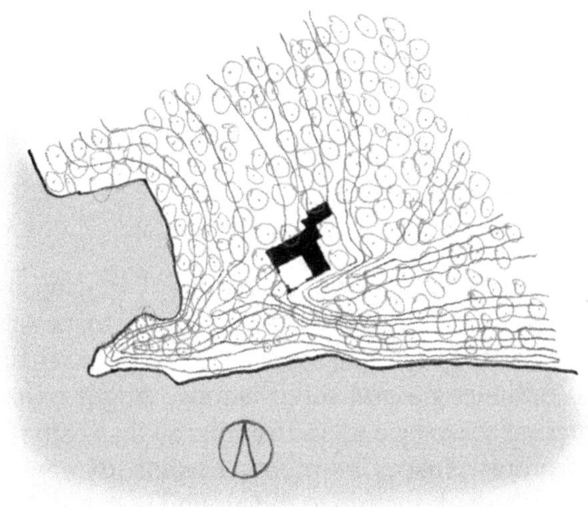

Figure 8.9 Site Plan of Alvar Aalto's Experimental House in Muuratsalo, Finland Is Nested in the Forest on Lake Kivijärvi in Central Finland; North Is Up as Arrow Indicates.

Figure 8.10 Alvar Aalto's Experimental House in Muuratsalo, View From Lake Kivijärvi in Central Finland; the House Is Elevated From the Lower Lake and Set Back in the Forest to Diffuse Winter Winds Blowing from the Northeast.

Geology

Broadly, geology is the study of the earth, soils, rocks and sediment structure. With regard to natural systems and thermal mitigation in architecture, we are generally concerned with ground temperature and perhaps the moisture content and constancy. Shallow average ground temperatures (nominally 5-ft depths) remain relatively constant annually varying only about /−10°F through summer and winter. Coupling architecture to the ground can be an effective natural system of reducing loads on the passive and active systems in buildings. The map below in Figure 8.10 illustrates the nominal annual ground temperatures which are much lower that average summer high and higher than the winter low temperatures. In cold climates, this is usually a strategy to reduce heating loads. In hot climates, it reduces the need for air conditioning or extends comfort further into the summer. However, in hot humid climates, the cool surfaces against hot humid air can cause condensation on building surfaces which may be unsightly or cause molds to form. In this case care must be taken to control humidity.

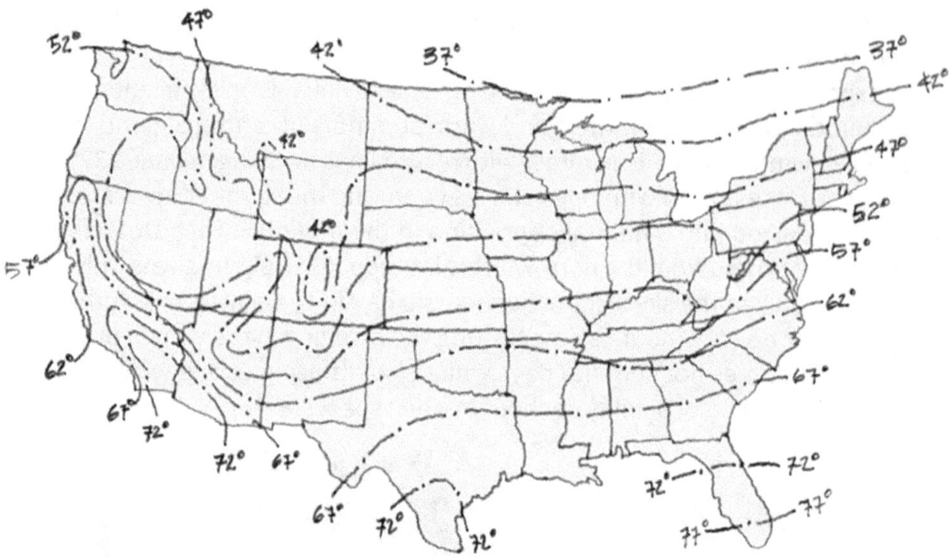

Figure 8.11 Shallow Ground Temperature Contours Across the United States Mainland in Degrees Fahrenheit; the Data Is Adapted From the United States Environmental Protection Agency—www3.epa.gov; Detailed Ground Temperatures May Be Available From Local Weather Stations.

Figure 8.12 William Morgan's Dickenson Hall, Part of the Florida Natural History Museum, Is Embedded Into the Earth, Along the North Façade (Top) While Opening to the South (Bottom).

Frank Lloyd Wright's Jacobs II House, also referred to as the Hemicycle House due to its semicircular shape in plan opens to the south seeking to optimize natural solar heating. Wright utilizes ground coupling and earth berms on the north side to protect spaces from winter winds. By embedding the architecture in the ground with temperatures warmer than winter air, the site is used as natural insulating system. The upper bedrooms have small horizontal windows to the north to allow daylight and cross ventilation in winter. This thin building form, coupled to the ground, and open to the south, seeks to optimally balance natural and passive design for the climate in Middleton, Wisconsin. There are many references to the house on the internet including this e-article: https://journals.open.tudelft.nl/dash/article/view/4720

Florida architect William (Bill) Morgan designed multiple earth-coupled homes, commercial buildings, and the Florida Natural History Museum, now Dickenson Hall, on the University of Florida campus. This natural strategy for hot humid climates takes advantage of the nominal 72°F ground temperature in contrast to the long, hot, North Florida summer with average daytime air temperatures well above 90°F from May to September along with peaks approaching 100°F. For daylight, Morgan opens the building to the south with long overhangs and extended balcony terraces that shade from direct sun and bounce sunlight into the occupied spaces. The images below are taken on cold winter morning (9:00 am January 28) showing the sun penetrating and warming the interior spaces. Morgan takes advantage of the natural slope of the site complimenting it with strategic earth shaping to form a terraced plaza as an extension of the building into the landscape.

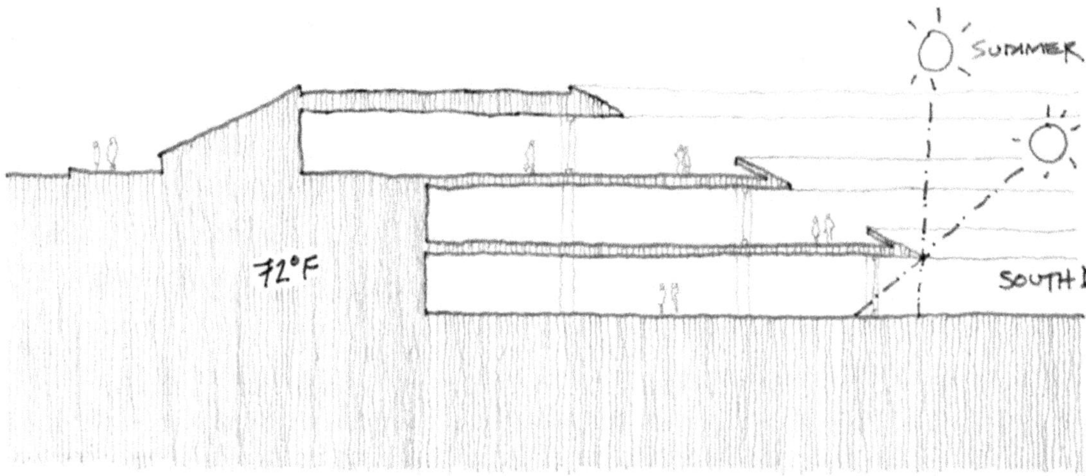

Figure 8.13 Section Sketch of Dickenson Hall Illustrating the Embedded Architecture with Consistent Annual Ground Temperature of 72°F.

Vegetation

Sites located in open plains or agricultural fields with wide open and flat spaces often contend with strong winds particularly in the winter. The openness offers unobstructed access to sunlight to be embraced in the winter, yet shade may be needed in the summer. Japanese farmers in Shimane Prefecture in Western Japan have nurtured tall dense walls of pine trees growing up to 50-ft high. The "L" and "U"-shaped wind breaks shield the dwellings from winter winds leaving open areas toward the south to catch sunlight. The climate can be snowy and windy in the winter months.

Careful consideration of the deciduous (loses leaves in winter) or evergreen trees (keeps leaves throughout the year) is another natural system strategy for mitigating harsh thermal environments. Evergreen trees are typically pines that have needles rather than broad leaves. Pines can still be very effective at providing shade in hot and hot humid climates. In the case of deciduous trees, they can be very effective in seasonal climates by providing shade in summer and allowing the sun to penetrate spaces during the winter.

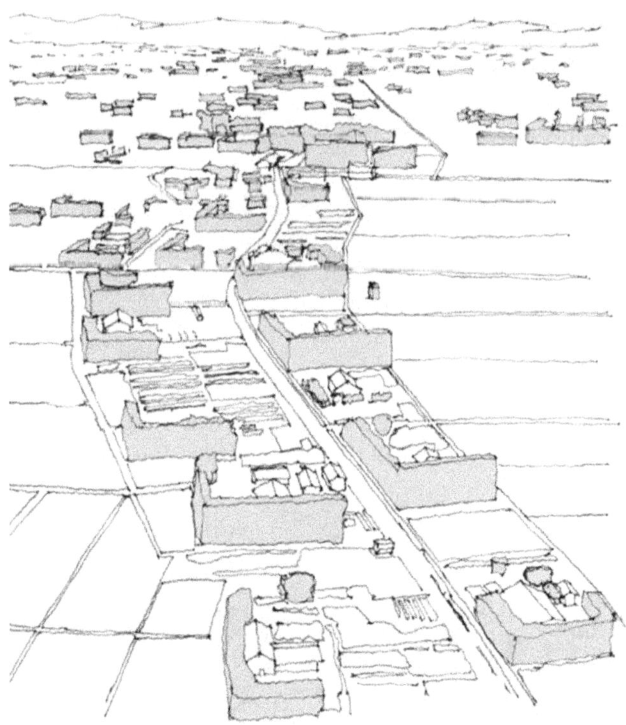

Figure 8.14 Vegetative Wind Breaks Around Dwellings in Shimane Prefecture in Western Japan (Adapted from Photographic Plate by Bernard Rudofsky, Architecture Without Architects, 1965).

Figure 8.15 Sketch Illustrating Shading/Warming Effect of Carefully Planted Deciduous Tree.

Figure 8.16 Highline, Manhattan New York, Deciduous Trees in Summer.

Figures 8.16 and 8.17 below show summer and winter conditions on Manhattan's now famous Highline linear public park. The deciduous trees provide much needed shade in the city's hot summer while allowing the sun to warm the walkways and benches in the winter. In the winter image (Figure 8.17) evergreen trees are interspersed at the edges of the path to provide limited shade but also to maintain the "green" park character during winter months.

Water

Oceans, seas, gulfs, and lakes can provide local thermal moderating effects within larger climate classifications as the high thermal capacity of water is slow to respond to air temperature changes resulting in much more stable temperatures than the air and even the surrounding earth. During the day, natural thermal columns of warm air rise over land (heated by the sun) and cooler air typically falls over water bodies. This phenomenon can also occur when land surfaces are thermally different and adjacent. For instance, large dark areas of soil (unplanted crop fields for example), or asphalt will cause the air over them to heat and rise in the sun while cooler ground surfaces such as lakes, wetlands, or densely vegetated areas will have sinking air columns above them. The convective air flow from water bodies can cool the adjacent air and can generate breezes along the ground near large water bodies.

Figure 8.17 Highline, Manhattan New York, Deciduous Trees in Winter.

Paul Rudolph's Bernhardt-Cohen house on Casey Key, Florida, is a great example of natural systems being used to offer comfort a somewhat harsh hot and humid location. At 27° NL, sunlight is plentiful and comes primarily from above during the midday but with the site so open to the east and west, low morning and afternoon overheating must be mitigated. Rudolph tucks the building into the natural oak grove providing shade in the morning. He orients the house facing west southwest to capture the breezes in the day and evening as they change directions across the narrow key. The narrow building form is elongated perpendicular to the breezes maximized ventilative cooling as a natural system response to the site. To mitigate the potential heat from the low west sun, Rudolph uses a sliding lattice system along with huge overhangs (Figure 8.19). The west facing lattice system is operable and can be completely or partially opened. A second layer of glass sliding doors, a later addition, can also be deployed if the breezes are too strong or the air is too cold. On the east side, sliding glass doors are used so the main living space can be completely open to breezes. Perhaps a lesson from the Indigenous Native American chickee.

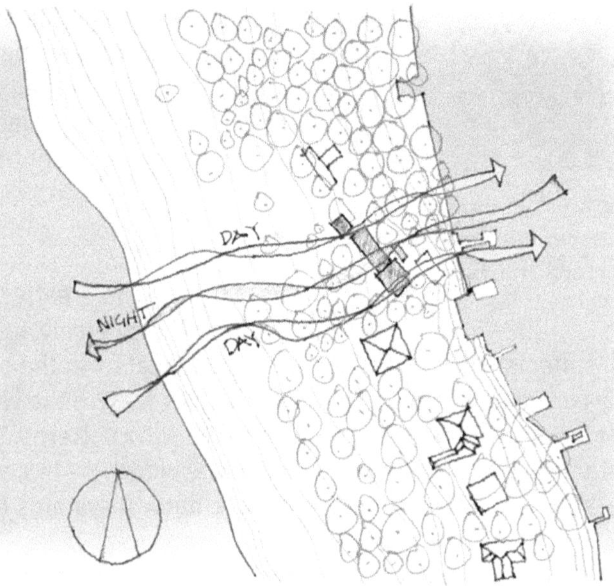

Figure 8.18 Bernhardt-Cohen House by Paul Rudolph Situates Itself in the Shade of Oak Trees, Presenting a Thin Bar to the Prevailing Breezes Maximizing the Natural Cooling Potential of the Site.

Figure 8.19 Bernhardt-Cohen House by Paul Rudolph Showing the Operable Lattice and Glass Panel System Utilized to: (1) Open Completely to Allow Full Breezed Through the Main Living Room; (2) Close the Lattice Screen to Provide Shade From the Low West Sun While Still Allowing Cooling Breezes Through the Space; and (3) Using Sliding Glass Doors to Retain Heat During the Short Cool Winters or Limit Breezes When too Strong.

Figure 8.20 Snow on Roof Can Act as an Insulator in Harsh Winter Climates—Low Slope or Flat Roofs Can Maintain the Snow Layer for Long Periods of Time.

As discussed with the Inuit Igloo above, snow can be a great insulator during the winter. Two schools of the thought exist regarding snow in modern structures. (1) shed the snow and ice as soon as possible by using steep roof pitches or by heating the underside of the roof. (2) keep the snow in place and accumulating to form an insulating barrier. Lighter colored roof surfaces (snow) radiate less heat to the cold night sky. Accumulation of snow is a natural strategy that can mitigate harsh cold environments but the structural load must be considered in addition to detailing the edges of the building to protect from ice damn damage or large blocks of ice and snow from falling on people or other property.

Breeze

Even a slight breeze can greatly extend comfort in hot humid environments and conversely, light winds in cold seasons can create significant wind chill effects and carry away substantial heat from the exterior surfaces of buildings. Breezes change seasonally commonly coming from the north or west in the winter and south or southeast in the summer; however, local conditions greatly influence the wind patterns. There are many internet sites that provide wind rose data from weather stations around the world. It's important to look at the seasonal changes rather than the just annual average to be sure the natural systems under consideration are seasonally appropriate.

Charleston, South Carolina is an excellent example of both urban organization and individual housing forms combining as a 'natural systems strategy' to capture the prevailing coastal breezes. The "Charleston Side Yard" leaves voids in the urban fabric that allow breezes to flow through the neighborhoods. Unlike the row houses of Boston, New York, or Washington DC, Charleston specifically leaves voids in the street for the breezes to flow through.

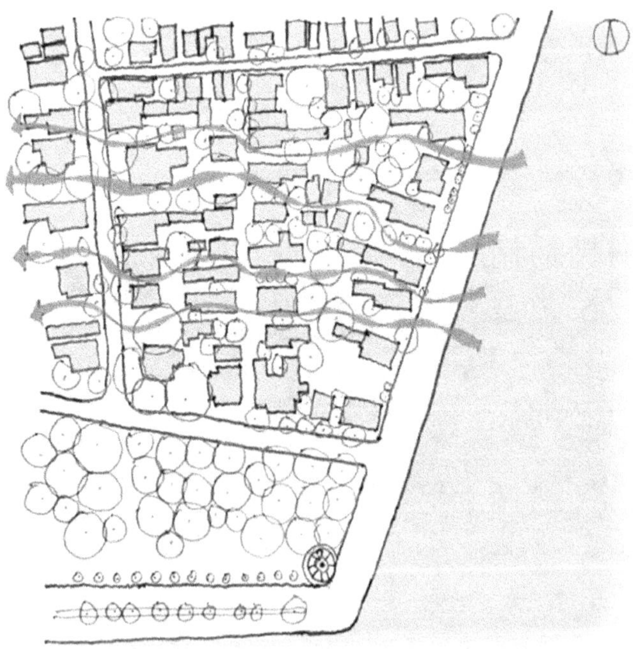

Figure 8.21 Charleston, South Carolina Urban Design Utilizing Side Yards to Allow Coastal Breezes to Penetrate Deeply Into the Urban Form (Sketch, Martin Gold 2023).

Visiting a traditional Charleston home, one would make formal entry from the street to the exterior side yard or garden, then to the south side porch. In summer, all the activity would take place on the porch as an intermediate inside/outside space. Even sleeping was done out on the porch during the hot summers. A common phrase at the time was to "mind your north side manners" referring to not looking out of your north facing windows and into the space of your south facing neighbor's porch and balcony spaces. The sketch below shows how breezes work, flowing through the side yards and homes. Note the bare Crape Myrtle, a deciduous tree, that shades in summer, and allows sunlight to warm in winter, a great selection for a climate that can be oppressively hot in the summer time but yet still be quite cold during the winter months.

Figure 8.22 Charleston Side Yard Configuration Allowing Breezes to Flow Through the Residence While Allowing Air Flow Through the City (Sketch, Martin Gold 2023).

Breezes have been discussed above as beneficial aspects of proximity to water and difficulties where lands are open and breezes may need to be buffered or mitigated. Many of these natural strategies are overlapping with passive design strategies that can be complimentary to the natural systems strategies outlined in this module. What has been presented is not a complete list and natural systems responses are evolving. The concepts presented here should provide students with a basic understanding, examples, and enough information to seek out and further evaluate case studies, examples, and take measure of sites and the potential natural systems responses that may be appropriate.

References

ASHRAE. (2021). *ASHRAE handbook—Fundamentals (I-P)*.

Bradshaw, V. (1993). *Building control systems* (2nd ed.). John Wiley & Sons, Inc.

DeKay, M., & Brown, G. Z. (2014). *Sun, wind, & light*. John Wiley & Sons, Inc.

Givoni, B. (1976). *Man climate & architecture* (2nd ed.). Van Nostrand Reinhold.

Google Earth Pro.

Grondzik, W. T., & Kwok, A. G. (2019). *Mechanical and electrical equipment for buildings* (13th ed.). Wiley.

Lindsey, R. (2009, January 14). *Climate and Earth's Energy Budget*. Nasa.gov; NASA Earth Observatory. https://earthobservatory.nasa.gov/features/EnergyBalance

Iowa Environmental Mesonet. (n.d.). *Wind Roses*. Iastate.edu. https://mesonet.agron.iastate.edu/sites/windrose.phtml?station=SVDG1&network=GA_DCP

Olgyay, V. (1992). *Design with climate*. Van Nostrand Reinhold.

Rodofsky, B. (1987). *Architecture without architects*. (Reprint edition.). University of New Mexico Press.

United States Environmental Protection Agency. jne_Henrys_map.html, www3.epa.gov.

Image Credits

Fig. 8.11: Adapted from "Shallow Ground Temperature Contours," https://archive.epa.gov/ceampubl/learn2model/part-two/onsite/ex/jne_henrys_map.html, 2017.

Fig. 8.20: Copyright © 2014 Pixabay/Simon Steinberger.

Module 9

Sun Path and Shade

Someone is sitting in the shade today because someone planted a tree a long time ago.

—Warren Buffet

Introduction

Figure 9.1 LeCorbusier's Juxtaposition of Aphrodite and Medusa Teaches Us That the Sun Can Be a Beautiful Warming Presence Brightening an Otherwise Dreary Day Yet Also Having the Intensity of Heat and Glare on a Hot Summer Day to Turn Us to Stone; The Paradoxical Power of the Sun to Warm Us or Bake Us, Must Be Well Understood by Architects.

Although we will be discussing sun and shade, two things about the title and quote above are somewhat misleading. Firstly, the Earth is making a path around the sun (earth path) rather than the sun traveling along a path, even though we refer to it as the "sun path." We refer to "sun path diagrams" and the sun "setting" in the evening and "rising" in the morning when in fact, we (people on earth) are setting and rising as the Earth spins and makes the sun visible to us each day. And, by traveling around the sun each year, the Earth carves out the seasons. Secondly, Warren Buffet—the Oracle of Omaha—was not really talking about shade and trees, but rather investment strategies—"be in it for the long-term, not the short-term." Yet, this is a perfect quote for us, as architects, who will provide shade for someone far into the future. Someone who might find that shade on a hot sunny day and perhaps enjoy a morning or afternoon. The task of making shade in architecture is not as simple as planting a tree, unfortunately, but planting a tree is still a good thing to do. Creating the sun and shade balance requires design sensibility. Shade and direct sun are essential to provide comfort, reduce energy use, and protect materials and people from the harsh radiation embedded in sunlight. With thoughtful design as a response to the sun/shade dichotomy, architects can draw out rich inspiration in seeking complex, textured, and meaningful architecture.

The Sun and Us

Sunlight heats the planet dynamically over the day-night and seasonal cycles. As the case will be made, all of the energy stored or utilized on Earth with the exception of geothermal and nuclear, is derived from the sun's energy, including coal, oil, natural gas, tidal, hydroelectric, and wind. It is quite fascinating that sunlight that was concentrated and stored millions of years ago can now heat our buildings and power our cars. However, there are costs to using fossil fuels—CO_2 emissions, among others. When the sun strikes surfaces on the ground or on our buildings, it can raise the surface temperature to over 180°F even when the air temperature is as cool as 65°F. If we are seeking heat, direct sun is a great benefit. However, if the air is 65°F to 100 °F, then this additional solar heat can be quite overwhelming. In this case, we must find shade.

Another way to characterize the intensity of solar energy is to establish a reference on how much energy we receive. In 1 hour, the sunlight striking the Earth's surface produces about 173,000 TWh (terawatt-hours or 10^{12} × watt-hours) of energy. In the calendar year 2023, the people on earth cumulatively used approximately 180,000 TWh for all the electrical production, transportation, and manufacturing conducted on the planet—basically all the energy consumed in an entire year. To summarize, the sun provides earth 13,000 TWh more energy in 1 hour than the people on earth consume in an entire year.

Turning to light, the sun offers abundant light during the day as a point source in the sky and as a diffuse source of light across the sky as the sunlight is scattered in the atmosphere under clear, partly cloudy, and overcast skies. Daylight is a combination of direct sunlight and diffuse light from the sky. We will investigate daylight strategies for architecture in detail in other modules. Here let's focus on embracing sunlight for warming spaces and strategies for shading to keep the sun from overheating our architecture.

Figure 9.2 McDonald's Headquarters and Restaurant Helsinki, Finland; Heikkinen and Komonen Architects Illustrating a Thoughtful Response to Shade, View, and Allowing Warmth from the Low Scandinavian Sun; Find It on Google Earth Pro at: Paciuksenkatu 29 00270 Helsinki Finland.

The cylindrical form of the McDonald's headquarters and restaurant in Helsinki, Finland, shown in Figure 9.2 perhaps responds to the nearly 24 hours of daylight occurring at just over 400 miles from the Arctic Circle. The shade screen extends from due east to due west on the south side of the building, pealing the low-angle sun into manageable slices, keeping the heat away from the building even in this predominantly heating climate. The density of horizontal slats that sieve and diffuse the light is offset against the windows of the occupied spaces suggesting a thorough understanding of how the shade would be projected onto the building from the angle of the sun. Pulling the screen away from the building keeps the heat at a distance, and it also acts as a windbreak during the winter, which is another important passive strategy during the windy Finnish winter. One might argue that this is contemporary adaption of Le Corbusier's "brise soleil"—deflecting heat while embracing daylight and airflow.

Sun and Sky Radiation

We can usually sense the heat from the direct sun on us, and we can feel that surfaces in the direct sunlight have been warmed or are even too hot to touch. On a partly cloudy day, clouds project shadows or shade to the ground and reflect direct sunlight back to outer space. On an overcast day, about 50% of the visible sunlight and 20% of the radiant heat is reflected to outer space. However, it is estimated that up to 80% of the ultraviolet (UV) radiation—the main heat source from the sun—can be transmitted through cloud layers as diffuse radiant energy. We can get a sunburn on an overcast day. This diffuse radiation in the atmosphere carries radiant heat that strikes buildings from all sides, not just from the location of the sun.

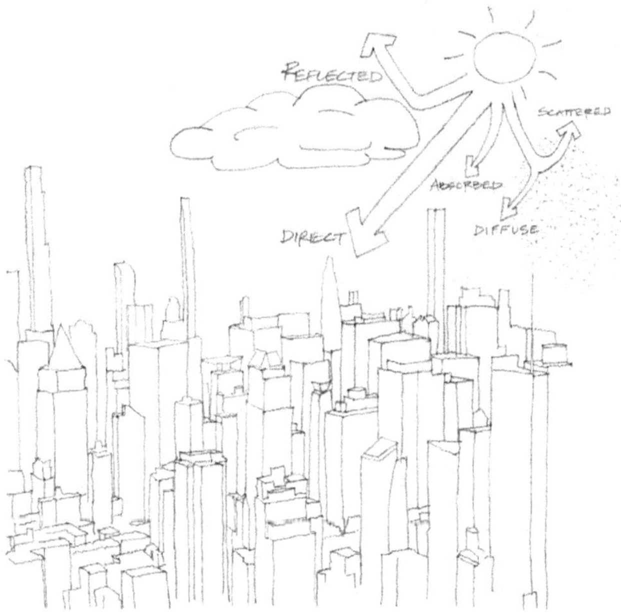

Figure 9.3 Sun and Sky Radiation Diagram Illustrating the Direct, Reflected, Absorbed, and Diffuse Radiant Energy Reaching Buildings in Midtown Manhattan.

Figure 9.4 Sky Conditions—Clear, Partly Cloudy, or Overcast—Affect the Amount of Radiant Solar Heat That Strikes the Surfaces of the Earth and Buildings.

Climate regions will have nominal annual or seasonal sky conditions of sunny, partly cloudy, or overcast periods. In the case of seasonal partly cloudy skies that perhaps form as part of the daily weather pattern, skies might be clear in the mornings and begin to cloud in the afternoon. In this case of partly cloudy skies, it is difficult to predict the amount of annual solar radiation that might impact a specific site. Rolling shadows from partly cloudy skies can reduce heat gain or daylight for short periods. Studies have also suggested that UV radiation can increase up to 25% on partly cloudy days. The mechanism for this is suggested that areas in the direct sun are receiving full radiant energy while nearby clouds are diffusing the solar energy, so at the location in the direct sun, it could be receiving direct sun in addition to diffuse UV radiation from nearby clouds.

Figure 9.5 Cumulus Clouds Casting Traveling Shadows Over an Agricultural Landscape.

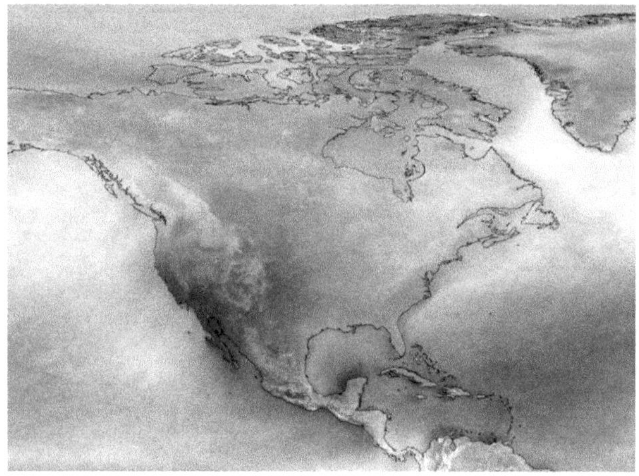

Figure 9.6 Cloud Cover Over North America Compiled for the Year 2018 (NOAA).

The National Oceanic and Atmospheric Administration (NOAA—pronounced like the name Noah) publishes a Global Data Explorer that allows users to select atmospheric data gathered from multiple locations around the world and to view the data in user-friendly maps: www.nnvl.noaa.gov/view/globaldata.html. The data in Figure 9.6 above shows cloud cover over North America as a summary for the year 2018. In the image, there are some clear distinctions in areas that annually receive more cloud cover. Figure 9.7 illustrates a summary of cloud cover data from NOAA compiled over multiple years. One can see that cloudiness is related to coastal proximity and geographic formations such as mountains.

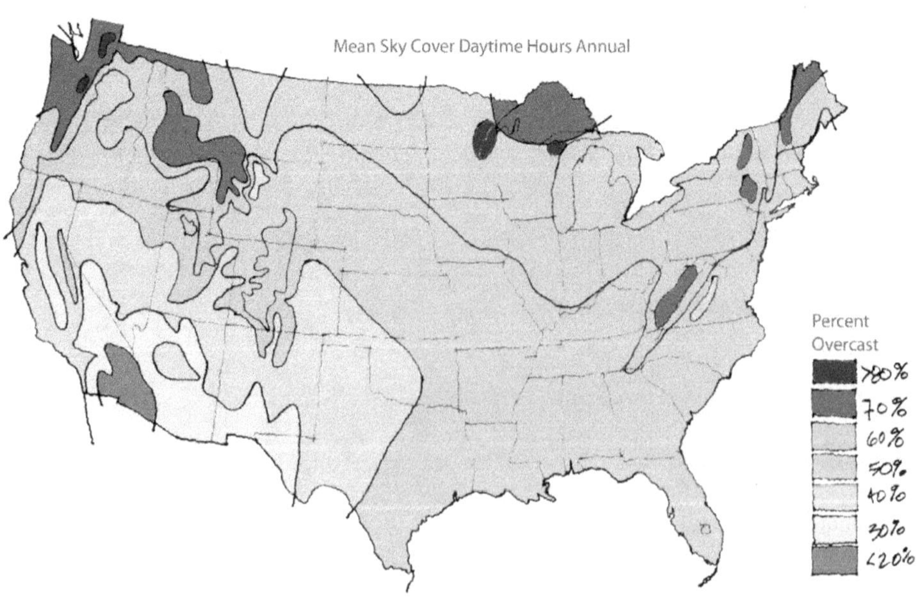

Figure 9.7 Percent of Daytime Hours with Overcast Skies USA (NOAA and NWS).

The conceptual illustration shown in Figure 9.8 illustrates the relationship between clear and partly cloudy skies in terms of radiant heat flow to the interior space. Both direct and diffuse sky radiation can penetrate glass as short-wave UV light heating the surfaces of interior spaces. That heat is absorbed and reradiated as infrared radiation (IR) with longer wavelengths than the UV light from the sun and sky. The long-wave infrared energy cannot escape through the glass as it is reflected back into the space and therefore, heat builds up in the space during the day.

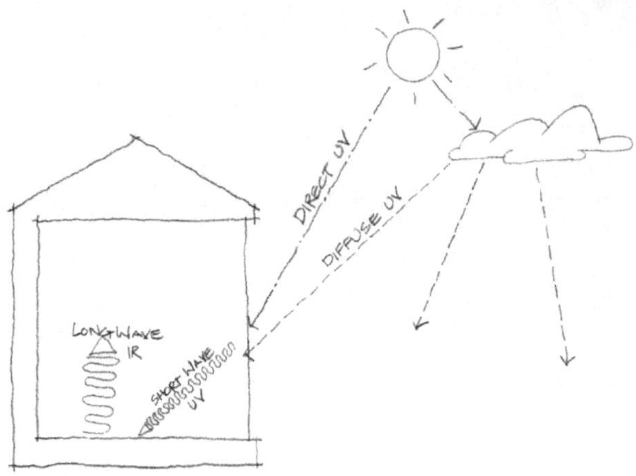

Figure 9.8 Diagram Illustrating the Effects of Direct and Diffuse Solar Radiation Impacting Architectural Glass.

There are many effective strategies to reduce the amount of heat that might enter spaces through windows, and in some cases, say in winter, one might want the heat to enter the space. In that case, careful design of the system or user operability for adjustment is required. The shading coefficient (SC) chart in Figure 9.9 above shows a range of opening systems considering the glass, interior, and exterior shading strategies. The shading coefficient rating is a ratio of solar gain that would penetrate the system relative to a standard 1/8-in thick float glass window.

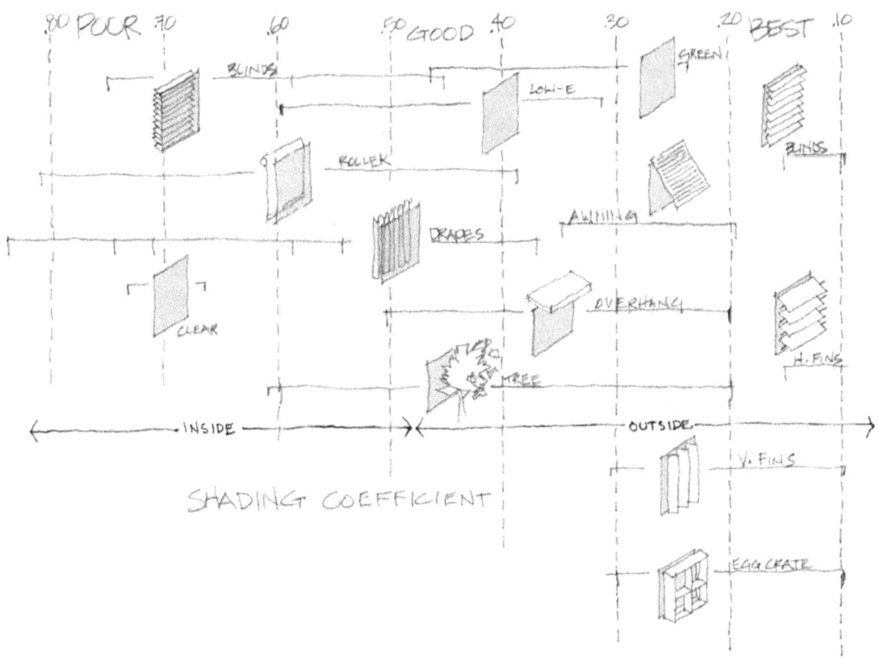

Figure 9.9 Chart of the Effectiveness of Window Systems Strategies to Provide Shade and Limit Heat Gain to Inside Spaces with Shading Coefficient Ratings.

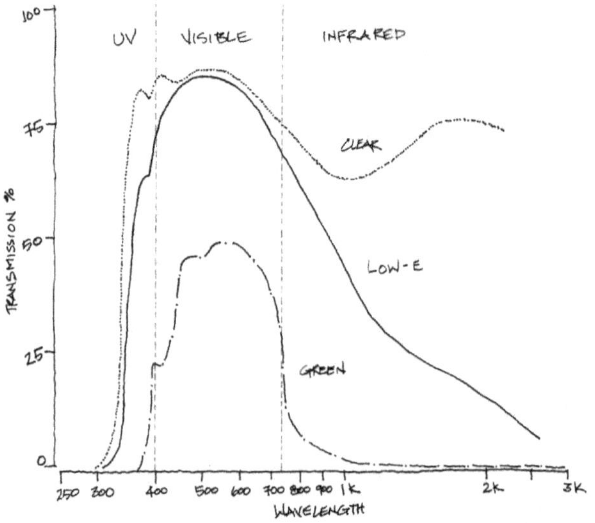

Figure 9.10 Graph Comparing Clear Glass, Low-E Glass, and Green Glass (Tinted Green); in Comparison, Green Glass Reduces Heat Gain the Most but Also Reduces Visibility Through the Glass and Limits the Amount of Visible Spectrum Light That Can Enter the Space.

Glass (glazing) itself can have a range of shading coefficients varying from poor to good when treated with tints or films. For example, low-E (low emissivity) glass is good at reflecting UV light and preventing it from passing through to the inside. However, if sunlight energy is desired for heating the space in the winter, then low-E would be a poor choice for the glazing. Many other glazing mix components or coatings can be added to tint or color the glass reducing the amount of the sun's UV energy transmitted through the glass. One drawback in using these coatings is that visibility is typically reduced and the effect on the visible spectrum must be considered in terms of viewing out or in as the case may warrant.

Mapping the Sun

Balancing daylight will reduce solar gain in summer and increase solar gain in winter which will contribute to healthful and resource-efficient architecture. If we consider shading and accepting heat from the sun as "passive design" strategies (no external energy source needed), then the design of fenestration (window systems) offers a great opportunity to balance light, heat, and shade. Passive design for fenestration is rooted in understanding the geometric relationship between the sun and the window system under consideration. The appropriate design depends on the latitude of the site, the orientation of the window, and any adjacent structures or vegetation that might block access to the sun. Various methods for determining the shading/heating seasonal cycles for windows have been developed, and most computer modeling software has shade/shadow modeling analysis elements embedded within the programs. To use those effectively, one needs to understand how that information is derived and the role of orientation and seasonal influences on the design. Additionally, shading strategies are not equally effective on each façade of a building. The following sections will introduce a process for mapping the sun path and adjacent skyline. Mapping the site sun path relationships can instill an intuitive sense of effective passive design responses to guide sun shading design.

Figure 9.11 Sun Over Meadow with Oak Tree Casing Shadows at 28°NL, February 25, 9:00 am.

In the sunny meadow shown in Figure 9.11, we can see the intense light from the direct sun but also get a sense of the diffuse light from the sky. Even with a clear blue sky (no clouds diffusing sunlight), the sun's light is diffused and where the sun is obscured by the tree, there is still light in the shadow. That light is the diffuse light (and heat) from the sky.

Researchers and scientists have prioritized solar altitude and azimuth as a method of placing the sun on a Cartesian grid and enabling the tracking of the path of the sun as viewed from a project site throughout the day and the year. The altitude and azimuth are used in various sun charting systems and provide a useful metric for designing shading devices such as fins, overhangs, and light shelves.

Solar altitude: The angular measure starting at the horizon and measuring upward (vertically) to the location of the sun. As shown in Figure 9.12 above, measuring up from the horizon to the sun, the altitude is 45°.

Solar azimuth: The angular measure toward the east or the west measured from due south. Solar azimuth always needs the designation east (E) or west (W) to identify the direction moving away from south. As shown in Figure 9.12 above, the azimuth, measuring toward the east, is 30°E.

In the northern hemisphere, when the solar azimuth is east of south, it is morning and when the solar azimuth is west of south, it is the afternoon. Students should nurture a good spatial understanding of the altitude and azimuth relationship of the sun to potential sites. For any site, establish true south early and keep the cardinal directions in mind.

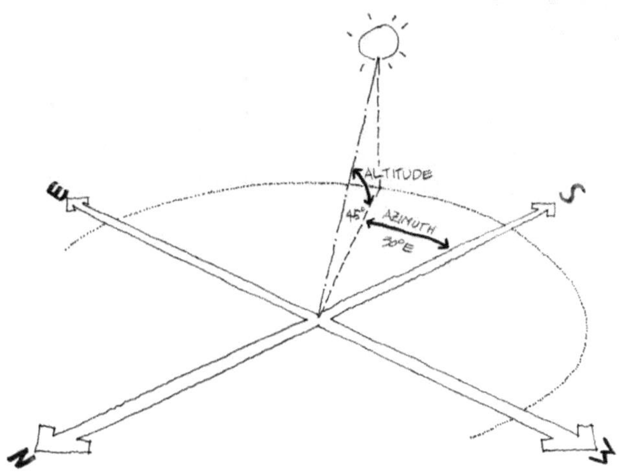

Figure 9.12 Illustration Indicating the Measurement of Solar Altitude and Solar Azimuth to Identify the Location of the Sun in the Sky.

With the understanding that we can mark the position of the sun in the sky with only two coordinates (altitude and azimuth), we can then track the sun's path in the sky over time, as viewed from a position on Earth, relative to a particular latitude, throughout a day, month, or year. Researchers during the late 1960s and early 1970s developed sun path diagrams to chart the sun's positions in the sky over the course of the year. Each sun path is only applicable for a particular range of latitudes (/− 10°). Latitude is the angular measure, in degrees of a location on the Earth's surface measured from the equator. For example, Miami, Florida, is at 25°NL (north latitude), Charleston, South Carolina, is at 33°NL, and Manhattan is at 41°NL. As one moves northward from the equator, the latitude increases until one arrives at the north pole—90°NL. The equator is 0° and north (NL) or south (SL) in the case of the southern hemisphere.

Figure 9.13 compares Victor Olgyay's and Edward Mazria's sun path diagrams. Olgyay's, on the left, is understood in plan, being viewed from above, through a sky dome that would be over a location at a specific latitude. Mazria's sun path diagram is taken from the vantage point of a person on the Earth, looking up at the sky, with the sky dome projected onto an orthogonal map. In either case, these are valuable tools to understand the position of the sun and to calculate the effectiveness of shading devices. For these modules, we will use the Mazria sun path diagram and shading calculator to evaluate shading strategies.

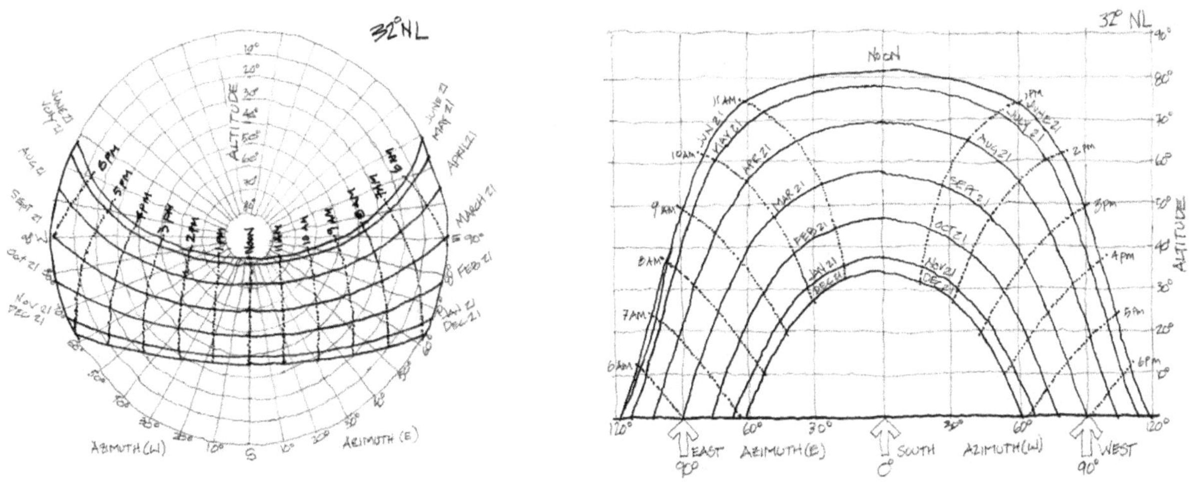

Figure 9.13 Comparison of Victory Olgyay and Edward Mazria Sun Path Diagrams.

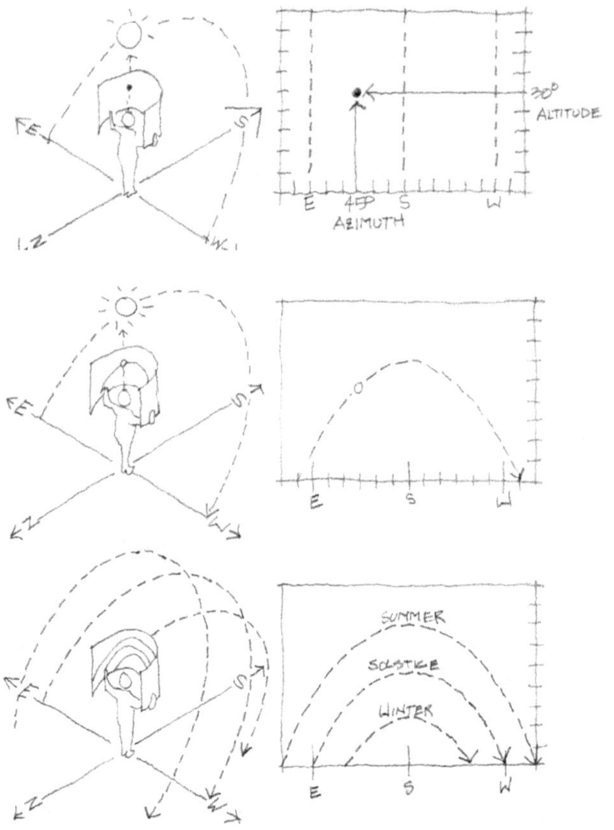

Figure 9.14 Edward Mazria Sky Dome Projected Onto a 2D Surface with Altitude and Azimuth Charting the Sun's Location at a Given Moment in Time (Top), the Sun Path on a Given Day (Middle), and the Seasonal Summer and Winter Equinox and Solstice Mapping out the Path of the Sun on Those Days for Each the Year.

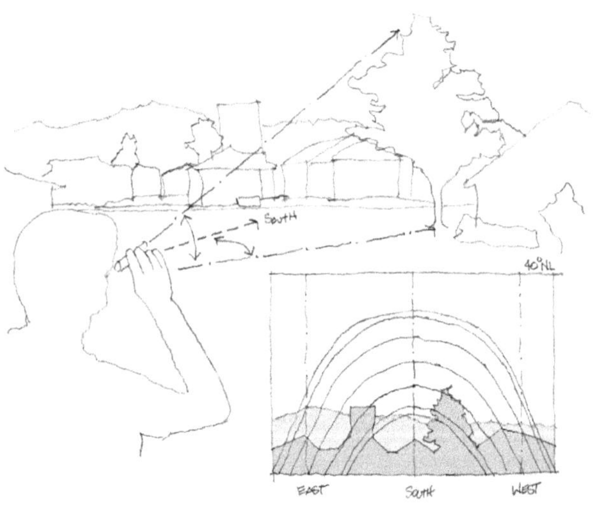

Figure 9.15 Mapping the Horizon on the Mazria Sun Path Diagram.

Within both sun path diagram systems, we can locate where the sun is for a given time of day and month. For example, if we knew it was 4 pm on August 21, the sun would be at 30° altitude and approximately 82°W azimuth. Note that with the Mazria sun path diagram, when we put it down on the desk, the west is on the right-hand side of the sheet which would normally be east if the north is oriented as up. This may seem a bit confusing so give this a try: hold a sheet of paper up to the south sky, mark south, east, and west, then put it down on your desk to see how west is now on the right-hand side of the paper.

From the diagrams in Figure 9.14, the methodology for mapping the sun with the Mazria sun path diagram should be clear. We can chart a moment in time (a point where the sun might be), or map the sun's path on a given day, month, or even throughout the year.

In Figure 9.15, we see the horizon being mapped onto the sun path diagram. This can be easily done for a location at a site with a simple, or more sophisticated, tool to measure the altitude angle of the environment around us—buildings, trees, mountains, etc. Figure 9.16 shows an example of a simple sextant that can be constructed using only some paper, a straw, tape, string, and a paperclip. If you don't have a straw, rolling up a piece of paper into a tube works just as well. With this simple device, you can measure the altitude angle of the objects and buildings at your local context or site.

Care needs to be taken to start the measurement looking due south, then marking the angular heights of things moving to the east or west and then finishing in the other direction. It is recommended to download a compass application for your smartphone allowing you to determine due south, and to become fairly accurate in the intermediate azimuth angles as when plotting the skyline on the sun path diagram. A physical compass will also work quite well if available. If using a physical compass one needs to understand magnetic declination. Magnetic north is not always the same as true north. When using a physical compass, a correction factor may be needed. If you use a smartphone app, you can usually set it to display either magnetic or true north. For an architectural design that relates to the position of the sun, we really want to know where true north is. There is a fascinating body of information on magnetic declination at https://en.wikipedia.org/wiki/Magnetic_declination.

The Mazria sun path diagram in Figure 9.17 maps the annual sun paths for 32°NL as indicated in the upper right-hand corner of the chart. This chart can be used anywhere on Earth that is along the 32°NL that circles the earth. For example, Charleston, South Carolina; Tucson, Arizona; San Diego, California; Shanghai, China; Tel-Aviv, Israel; and Tripoli, Libya, are all at about 32°NL, and the sun path diagram below would work in all those locations. Sun path diagrams for other latitudes are included in the appendix of this module.

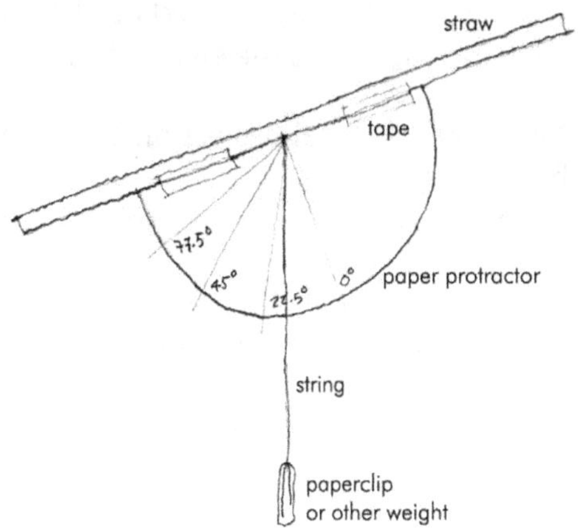

Figure 9.16 Simplified Sextant for Measuring the Azimuth of Environmental Elements Around a Particular Location.

The sun path diagram shows the sun's path for each month on the 21st day. Interpolation is needed for dates between the paths shown. Notice for example, March 21 and September 21 are indicated on the same path. This is because the sun is on that same trajectory twice each year. December and June only have one transit as those are the longest and shortest days of the year respectively—the solstices. The hours of the days are also indicated as dashed lines showing daytime hours from sunrise to sunset. In December, the sun sets just before 5:00 pm indicated by the path ending at the x-axis. The x-axis represents the horizon. Remember from Figure 9.15 that the sun may set earlier behind mountains or other buildings. The sun path diagram represents a clear open horizon such as near an ocean, large lake, or flat prairie.

Using the sun path diagram, take a moment to chart where the sun would be right now, as you are reading this (for 32°NL) or at some time tomorrow if you reading this late at night. What time would the sun set today/tomorrow at 32°NL latitude? What is the azimuth and altitude at 9:15 am on September 21?

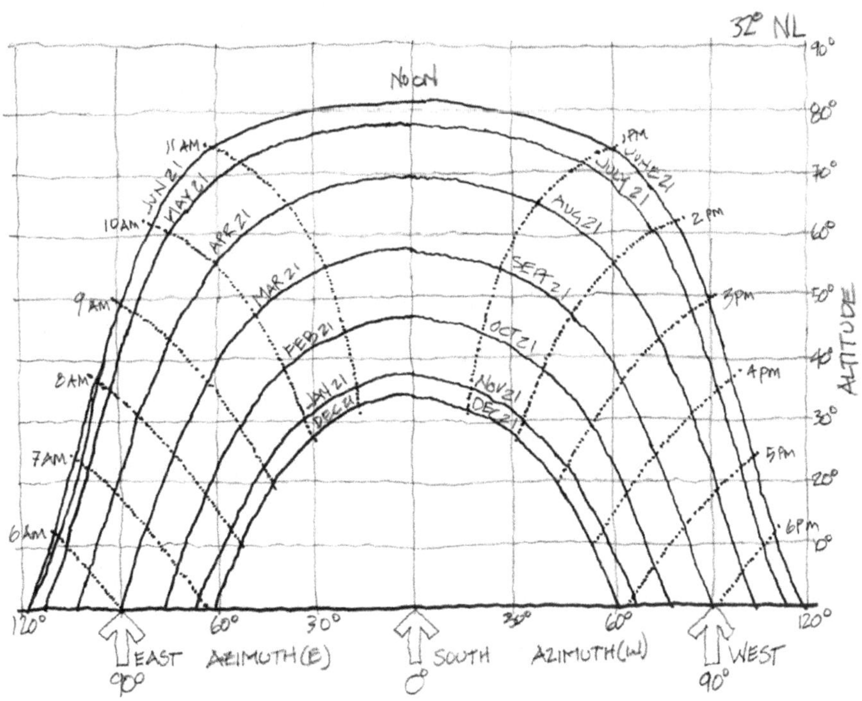

Figure 9.17 Mazria Sun Path Diagram.

Finding Shade

Now that we understand how to use the sun path diagram to map the sun's position, we can use it to design shading devices for openings (glazed or just open) in our buildings. First, we need to define a few terms. Figure 9.18 below illustrates the elements defined below:

- *Overhang:* a horizontal projection of the roof or shading device that extends outside past the wall of the building.
- *Fin:* a vertical projection extending outside past the outside wall of the building.
- *Cutoff Angle:* The angle extending from a location on the window to the far edge of an overhang or fin.

Overhangs and fins are the fundamental devices used for shading. There are numerous variations on the overhang theme and one might even consider louvers as a series of overhangs or horizontal shading devices. At the end of this module, multiple architectural examples and variations of shading devices are introduced.

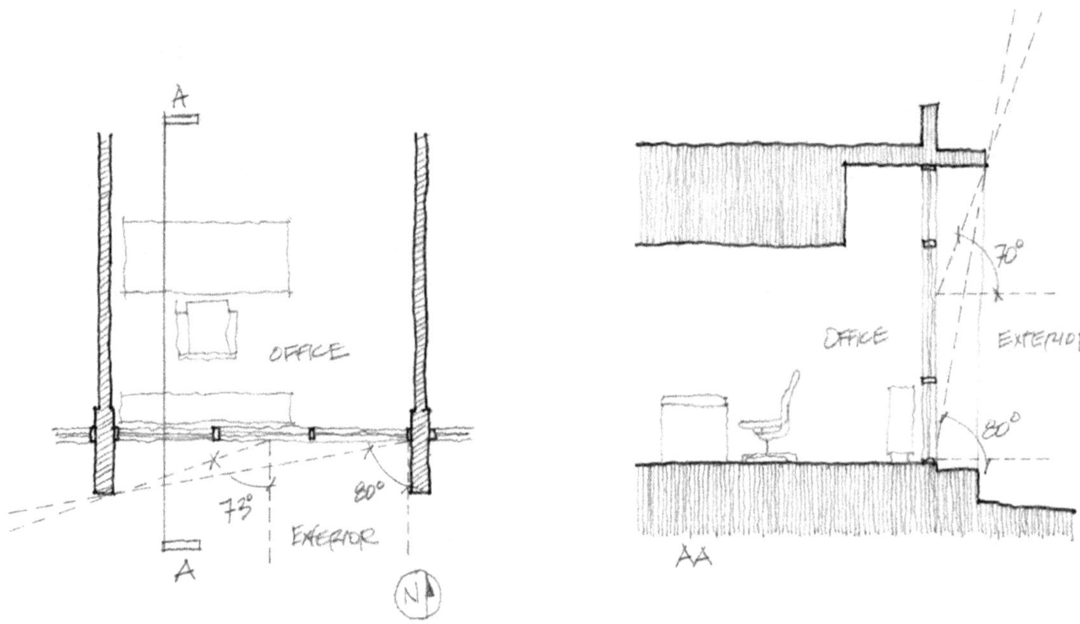

Figure 9.18 Illustration of Fin Shading (Left) and Overhang Shading (Right) with Associated Cutoff Angles.

The illustration shown in Figure 9.18 illustrates a plan (left) and section (right) sketch of an office with exterior overhang and fins. Cutoff angles are shown for 100% and 50% shading. The sun would have to be at (or greater) than 80° to provide 100% shade or at or greater than 70° to achieve 50% shading or more. The cutoff angle is critically important in evaluating the ability of an overhang or fin to shade an opening. It is also the angle that allows the sun path diagram and the shading calculator to be used to design appropriate shading devices and to test conceptual designs for ample shading. An overhang that is well designed can shade the window in summer when sun altitude angles are high and yet still allow the sun to penetrate into spaces to warm them during winter when the sun is low in the sky.

The shading calculator is a handy overlay tool used to measure the effectiveness of overhangs and fins and as a design tool to achieve specific shading strategies. The curved arcing lines on the shading calculator correspond to the shade provided by an overhang or other horizontal shading device. The vertical lines on the chart correspond to the shading provided by fins. To use the shading calculator, use the Portable Network Graphics (PNG) file version to use digitally or print the file onto clear acetate and overlay it over the sun path diagram. The angles on the shading calculator correspond to the cutoff angles of overhangs or fins. The arrow at the bottom of the shading calculator must be aligned to the azimuth angle on the sun path diagram to correspond to the direction that the window is facing. For example, if the window being studied is facing south, then place the shading calculator arrow at 0° (south). If the window faces 35°W of south, then place the shading calculator arrow at 35°W azimuth when overlaying the shading calculator on the sun path diagram.

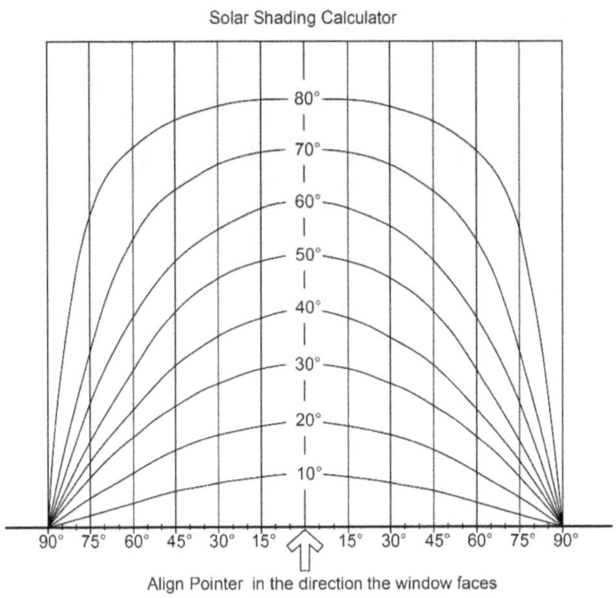

Figure 9.19 Shading Calculator to Be Used with the Mazria Sun Path Diagram.

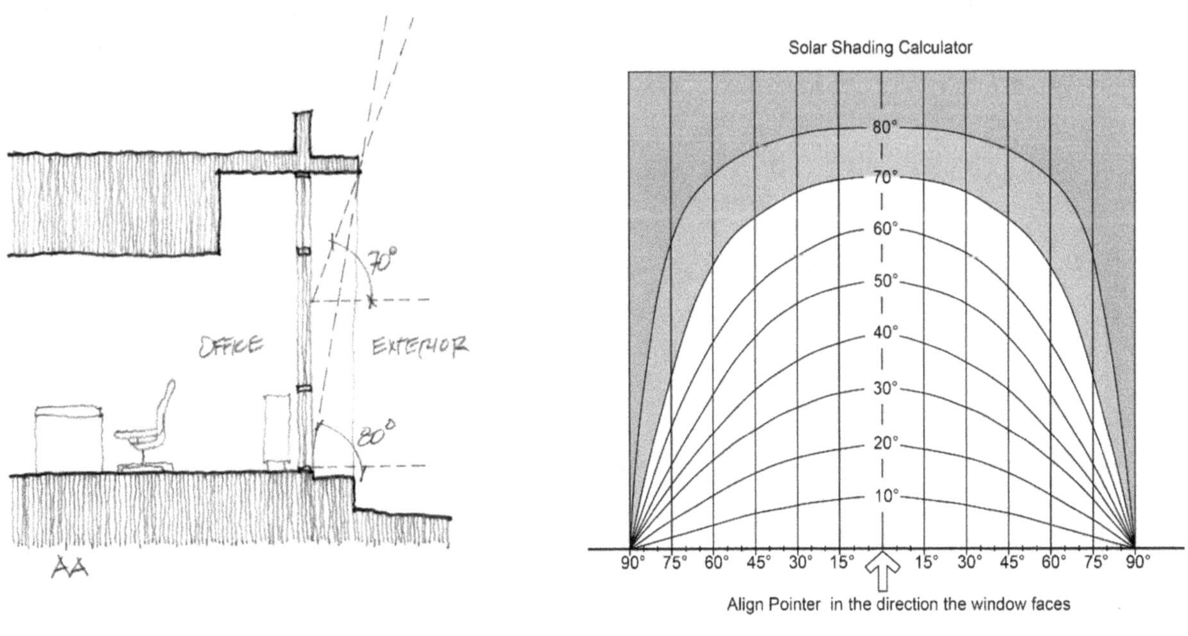

Figure 9.20 Example of Overhang for South-Facing Window.

Figure 9.20 shows both the 100% and 50% shading that corresponds to the 80° and 70° cutoff angles, respectively. When plotting the shade, the "shaded" area shown in orange on the diagram indicates the periods of time the window will be in shade.

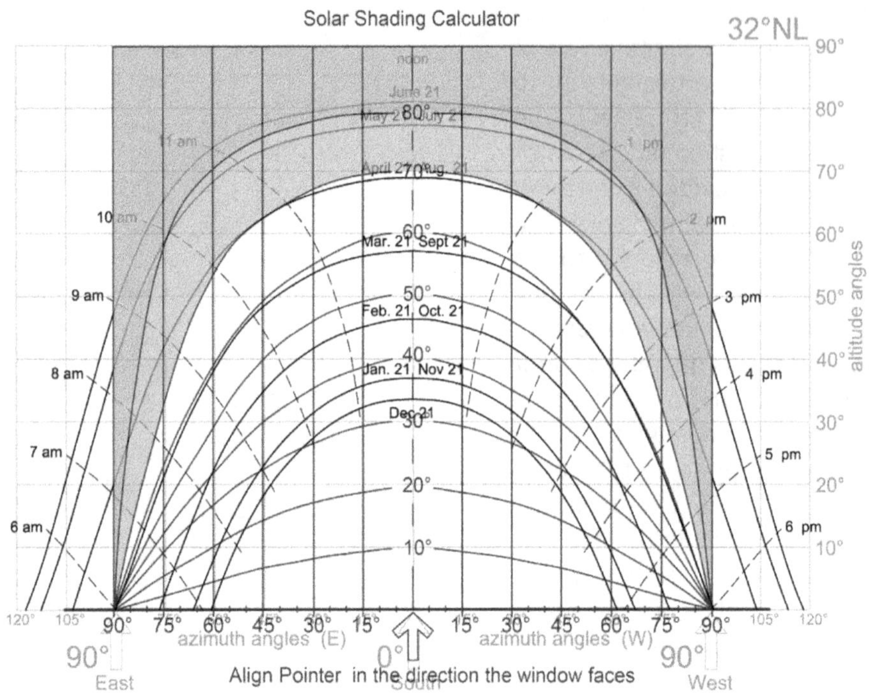

Figure 9.21 Overhang of South-Facing Window Plotted on Sun Path Diagram.

When we overlay the shading calculator on the sun path diagram, we see when the window is in shade (100% or 50%) based on the overhang condition that we diagrammed. Between April 21 and August 21, the window will be in 50% shade or better. However, the window will only be in full shade for approximately the month of June. During the times that are not covered with the orange tint, the window will be less than 50% shaded.

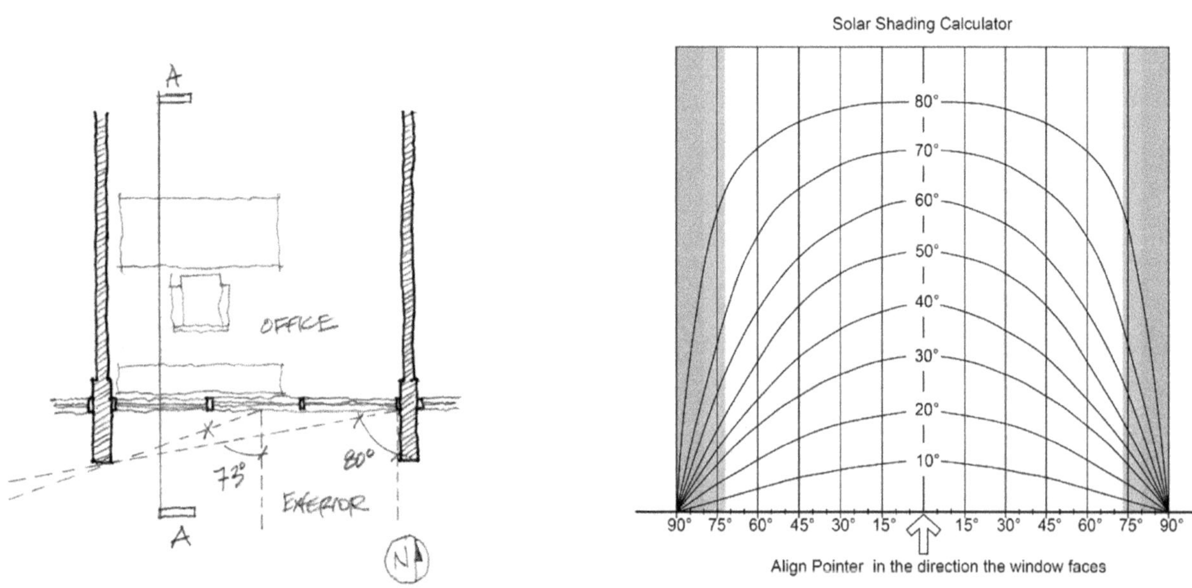

Figure 9.22 Example of Fin Condition at a South-Facing Window.

Figure 9.22 diagrams the effects of the fins on both sides of the windows. Fins are plotted on the azimuth angles. Not that the shade provided is always indicated away from 0° toward the outside of the solar calculator as shown. Both the 80° cutoff angle (100% shade) and 73° cutoff angle are represented with tints of orange.

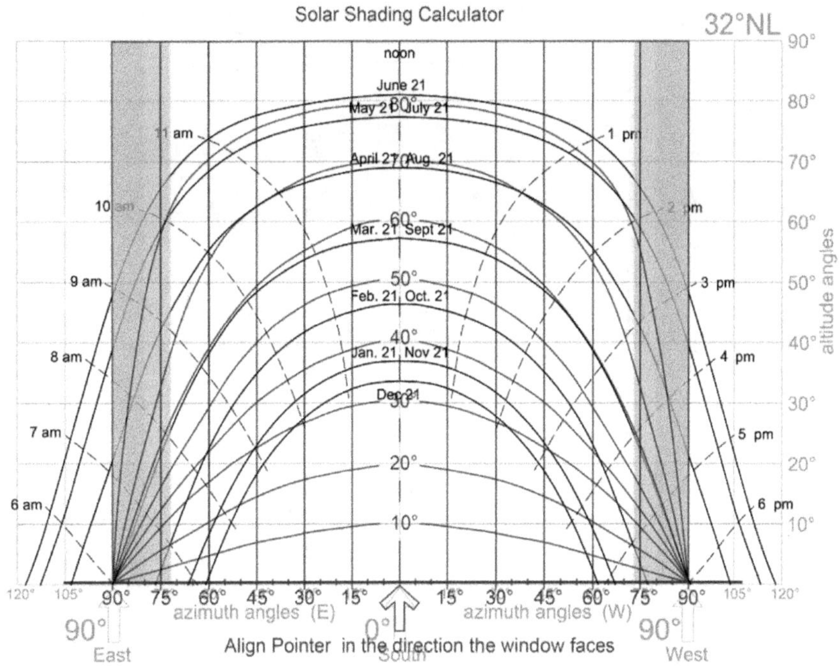

Figure 9.23 Fins from South-Facing Window Are Plotted on the Sun Path Diagram.

Plotting with the shading calculator as an overlay on the sun path diagram, we can see that the fins provide shade early and late in the day. The fins shade when the sun is low in the sky which happens closer to sunrise and sunset. This complements the overhang that works to shade during the mid day when the sun is highest in the sky.

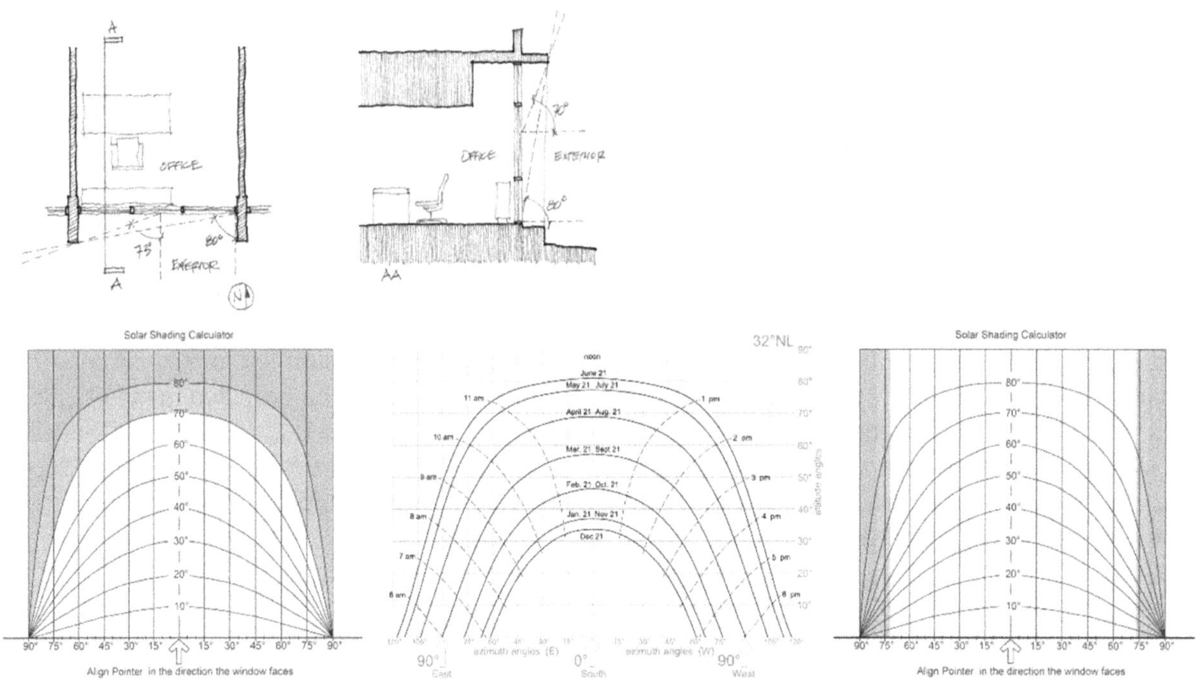

Figure 9.24 Overhang and Fin Shading Calculator with Sun Path Diagram.

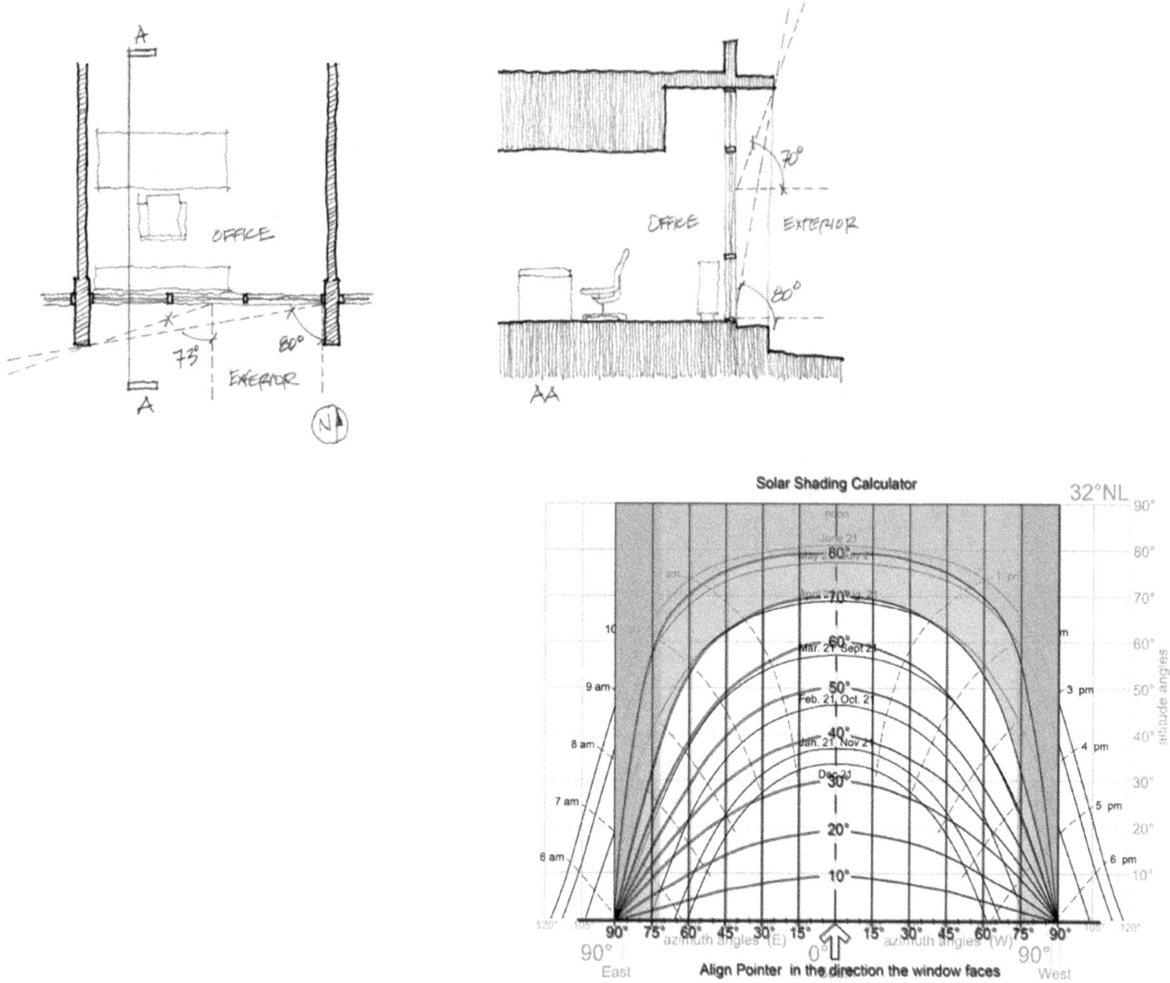

Figure 9.25 Composite Overlay of Shading Calculator and Sun Path Diagram for Fins and Overhang.

We can summarize the effect of the fins and overhang by plotting them together and referencing the plan and section of the architectural scenario—an office space facing south with moderate fins and overhang. Figure 9.25 combines the fins and overhang overlays to show an annual measurement of when the window is in full or 50% shade. There are periods when both the fins and the overhang are providing shade. Also, beyond the 90° angles (due east and west) the sun is not shining on the south-facing façade so at those times of day—early in the morning and late in the evening—the south-facing façade will not receive any sun. What is of some concern, is that during much of the early and late summer, the sun is penetrates deep into the space. At 32°NL, typically a warm latitude, more shade should be provided to maintain comfort in the space and reduce the load on mechanical equipment.

What if we were to extend the overhang to provide shade from March 21 through September 21. From the sun path diagram, we can determine we would need a cutoff angle of approximately 58°. Going back to the original design drawings, we can plot the 58° cutoff angle on the section drawing and then extend the overhang to meet that angle. Although the scale is not shown, this would require an additional 3 ft of overhang. Figures 9.26 and 9.27 illustrate the shade provided by the extended overhang with the 58° cutoff angle. Another approach would be to add a second overhang or several overhangs that do not project as far away from the building at intermediate heights across the window. It is important to note that with this option, the window could be shaded for longer periods of time, but the view from the interior could be compromised if the overhangs are in the viewers line of sight.

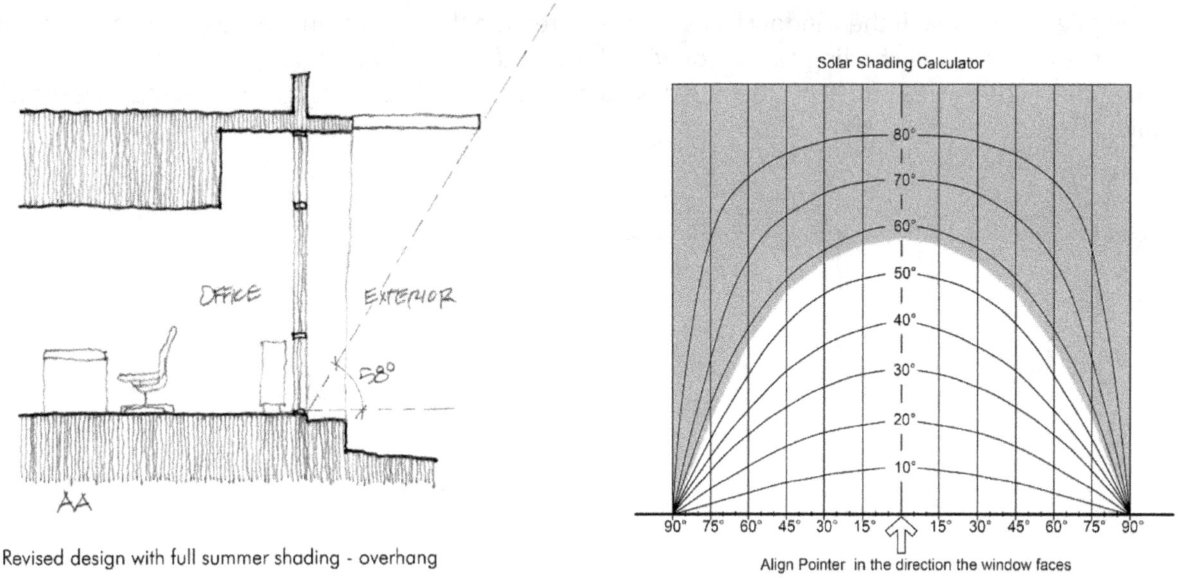

Figure 9.26 Example of Extending an Overhang to Provide 100% for a Longer Period of the Summer Months.

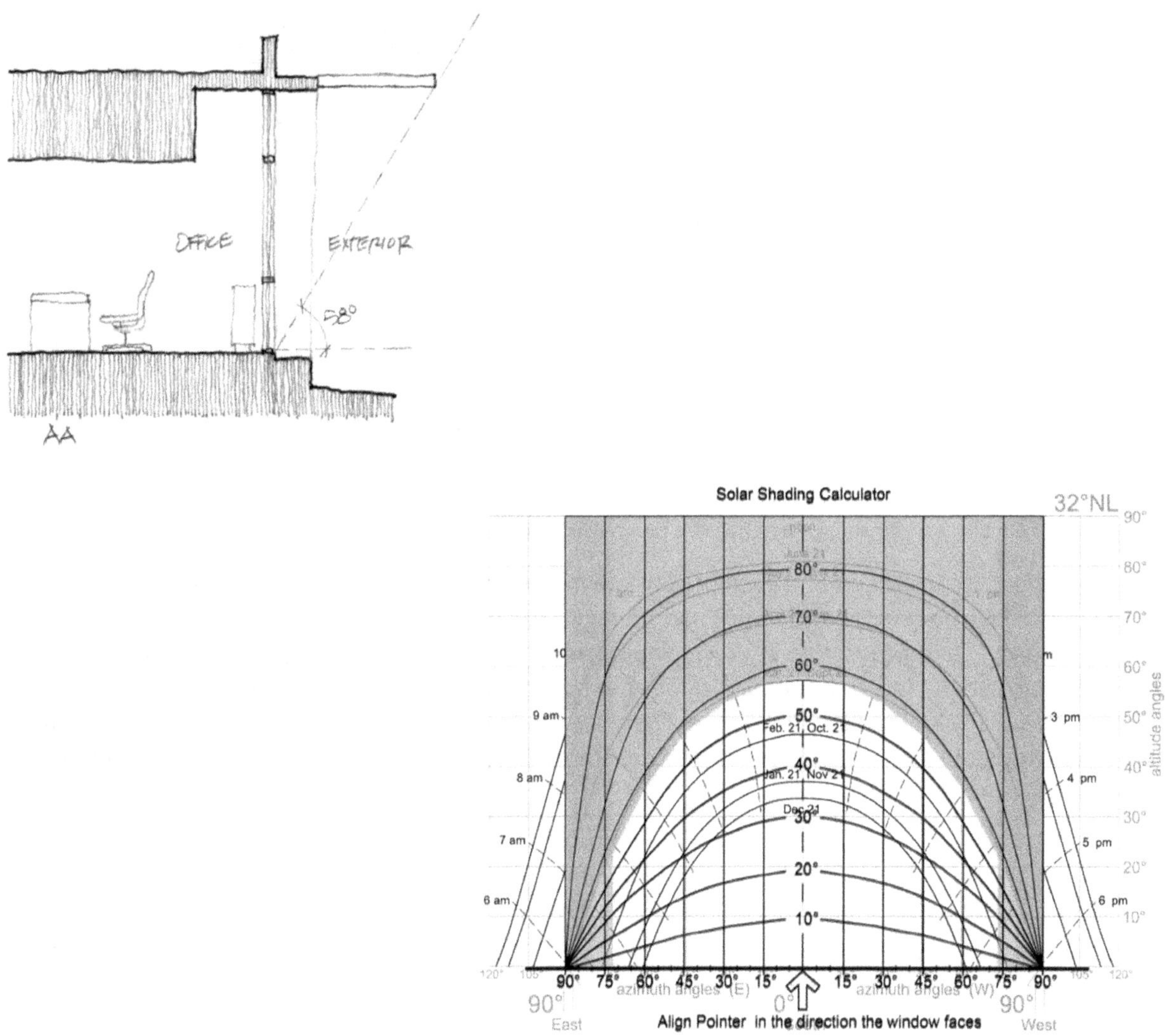

Figure 9.27 Extended Overhang with Existing Fins Plotted on Sun Path Diagram.

One might ask, what if the window faces another direction than south? In that case, one needs to simply align the arrow of the shading calculator with the azimuth direction the window is facing on the sun path diagram. Figures 9.28 and 9.29 illustrate this process for a southeast-facing window and a west-facing window.

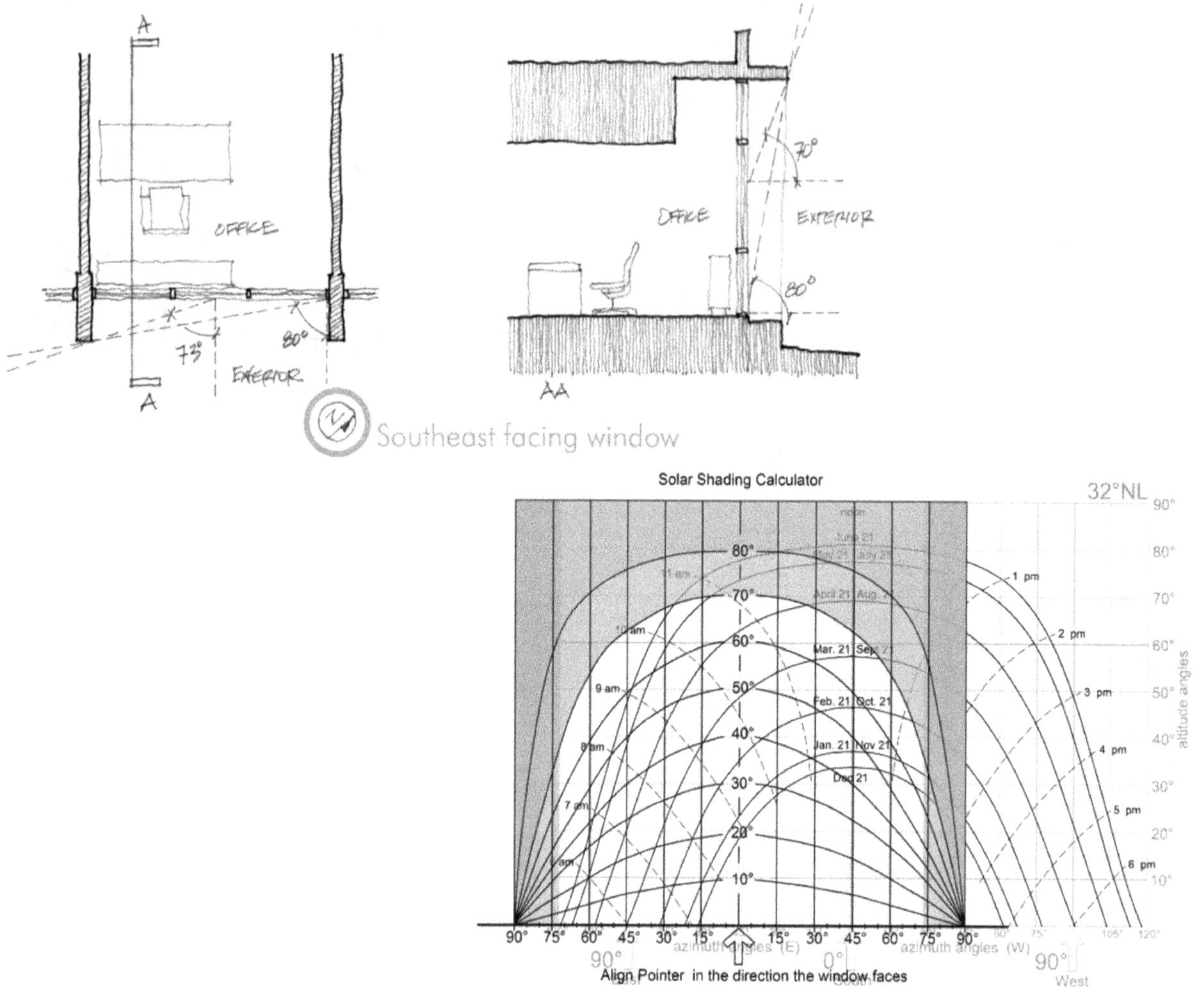

Figure 9.28 Example of an Overhang Condition at a Southeast-Facing Window.

Notice how the overhang and even the fins are much less effective as the façade faces more to the east or west. Even if we have extended the 58° cutoff angle of the overhang which is quite long relatively speaking. It is clear we cannot provide effective shading on the east and west façades with overhangs alone. Therefore, it is usually recommended to put buffer spaces on those façades (garage, storage area, or bathroom) that typically have less window area. There are other strategies, but east- and west-facing spaces are difficult to moderate summer and winter sun. South-facing spaces are much easier to work with to exploit potential solar opportunities.

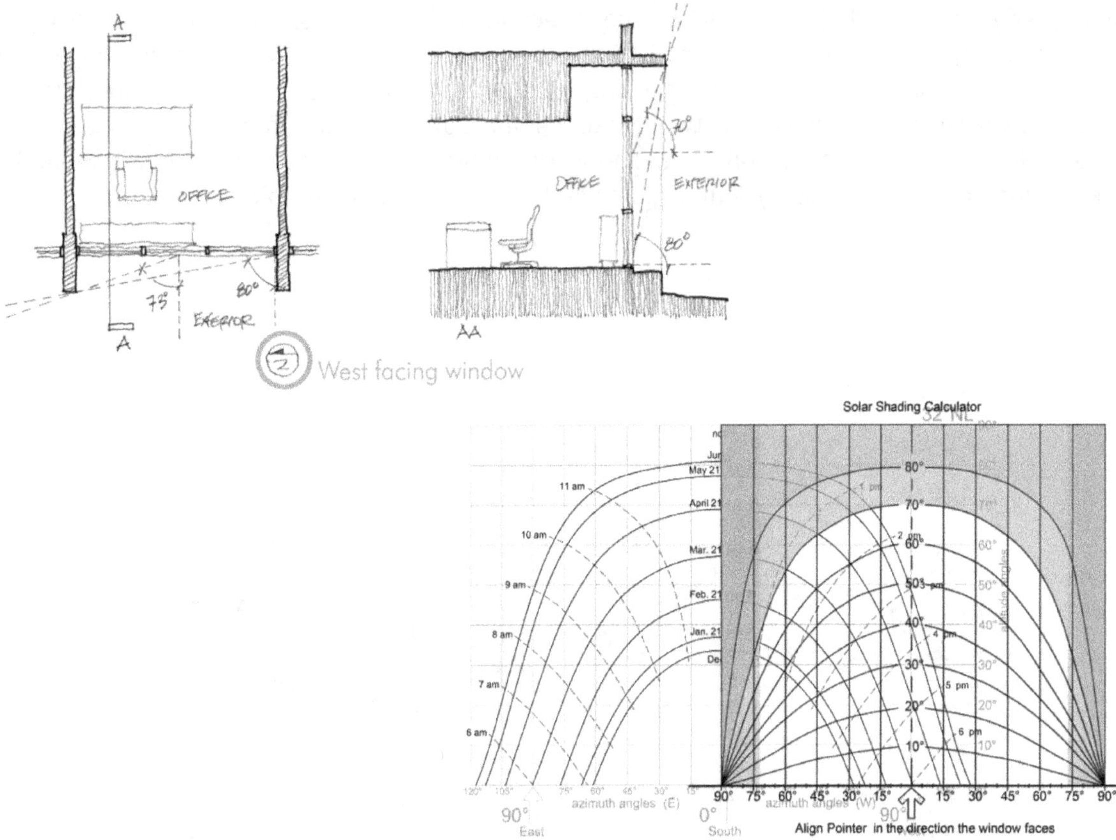

Figure 9.29 Example of an Overhang Condition at a West-Facing Window.

Architectural Responses

Selected projects are included below that illustrate the fundamentals of the overhang, fin, and shade or screen strategies for sun shading. These strategies are implemented within the context of the local environment, pay attention to view and daylight, implement design concepts that drive the overall form, consider the spatial organization of program, and draw from the materiality, and technical systems available. It is hoped that through this range of projects, these strategies will become intuitive and motivate further research to expand your skills toward implementing creative, playful, and inspiring well-shaded architecture.

Figure 9.30 Architecture Office of Gene Leedy, Gene Leedy, Winter Haven Florida; Office Entry Facing South.

Figure 9.31 Architecture Office of Gene Leedy, Gene Leedy, Winter Haven Florida; East and North Façades.

Gene Leedy was a noted "Sarasota School" architect practicing in Winter Haven, Florida, with multiple award-winning projects around the State of Florida. He was noted for his unique designs using structural "T" members in his residential and commercial projects that are typically reserved for transportation infrastructure. The structural spanning capacity of these members allowed Gene to make heroic cantilevers as primary shading devices. Gene also takes advantage of the large oak tree to the south of the building to increase the shading of his two-story sliding glass window system in a hot humid climate.

Figure 9.32 Burkhardt-Cohen Residence by Paul Rudolph, Casey Key, Florida—North and West Façades.

Architect Paul Rudolph orients this home with the long axis running from the south to the north due to the unique site between the Gulf of Mexico and intercoastal waterway. On this narrow key, views are optimal toward the east and west. Figure 9.32 shows the short south façade and the long west façade. Here you can see the heroic overhang, perhaps endemic to the Sarasota School architects (Rudolph being a leading member). The short overhang to the south with plenty of glass can provide ample shade while the west-facing windows need much more overhang to provide afternoon shade. The long west-facing opens to the west and Rudolph uses a system of sliding screens to either fully open the space in the morning and mid-day. As the sun transits to be low in the west the open lattice screen can be closed to provide shade while maintaining cooling breezes. Originally, there was no mechanical air conditioning in the residence.

Figure 9.33 Burkhardt-Cohen Residence by Paul Rudolph, Casey Key, Florida—West Façade.

Figure 9.34 Burkhardt-Cohen Residence by Paul Rudolph, Casey Key, Florida—Shade Screen.

Figure 9.35 Umbrella House, Paul Rudolph, Lido Key, Sarasota; North Façade and Entry.

Figure 9.36 Umbrella House, Paul Rudolph, Lido Key, Sarasota; Looking Northwest at the Umbrella Canopy and South Façade.

Figure 9.37 Umbrella House, Paul Rudolph, Lido Key, Sarasota; Looking Through the Jalousie Windows to the Pool Under the Umbrella.

When the Sarasota School architects were designing and building, homes in the area, they typically did not have air conditioning. Careful shading strategies allowed natural breezes to cool the spaces as an essential part of the design language in the mid 20st Century. In Paul Rudolph's Umbrella House—a developer spec house—Rudolph used jalousie windows (small slat glass planks that hinged open or closed) allowing the entire façade to be open to breezes. These jalousie window units were widely available and inexpensive from the Sears, Roebuck & Co. catalog at the time. In this case, the "umbrella" or shade screen, completely covers the home and southern terrace with a sunspace opening over the pool to control shade accommodating sun bathing on the terrace.

Figure 9.38 Cal-Trans Administrative Headquarters, Los Angeles, California, Thom Mayne (Morphosis); West Façade and Public Plaza.

In his award-winning design proposal for the new California Transportation Department (Cal-Trans) administration offices in Los Angeles, architect Thom Mayne was presented with a site requiring massive east- and west-facing façades. Here he used a double skin system by wrapping a traditional glass office block with a perforated "exo-skin" to shade the building from the low east and west sun. Operable shade screens over the windows are closed in Figure 9.38 with the sun bearing down on the west façade. In Figure 9.39, the many window screens are open operating as overhangs. As the sun transits the sky, the shades open to provide view and additional daylight when the sun is high in the sky while closing when the sun is lower in the sky. This prevents direct sun from penetrating deep into the spaces. The south façade is a system of fixed photovoltaic panels designed to provide shade at the south-facing windows. The building's shade screen and exterior stair system extends past the south façade acting as shade fins at the scale of the entire building. The same folding exo-skin is used for canopies to shade the public space and to shade and screen from view the mechanical cooling equipment on top of a lower portion of the building. Mayne seamlessly integrates sun shading, view, concealment, and inspiring design while meeting California's state budget for administrative facilities.

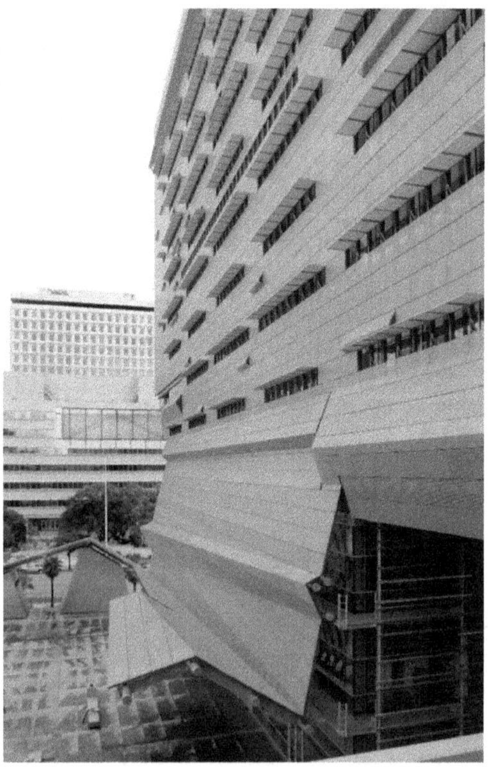

Figure 9.39 Cal-Trans Administrative Headquarters, Los Angeles, California, Thom Mayne (Morphosis); West Façade with Window Overhangs and Large "Unfolding" Overhang at the Plaza.

Figure 9.40 Cal-Trans Administrative Headquarters, Los Angeles, California, Thom Mayne (Morphosis); South Façade with Photovoltaic Array as Overhang/Light Shelf Shading Strategy.

Figure 9.41 Barcelona Pavilion, Ludwig Mies van der Rohe, Barcelona, Spain. Main Entry.

Figure 9.42 Barcelona Pavilion, Ludwig Mies van der Rohe, Barcelona, Spain. Reflecting POOL, LIBRARY, and Gift SHOP.

Mies's famous Barcelona Pavilion is perhaps a seminal example of the "floating plane" as an overhang. Marking the entry with protection from rain and sun, the overhang extends far beyond the interior space to provide shade from the southern sun for both interior and exterior spaces. The red curtain can be drawn in the morning to shield from the low morning sun in the east. The motif of overhang as entry is used again at the library and gift shop across the reflecting pool as a repetitive element reinforcing the overhang as entry idea.

Figure 9.43 Fondazione Feltrinelli, Herzog and de Meuron, Milan, Italy Shows the Architects Use of Extended Floor Slabs as a Shading Device Along with Deep Coffered Windows to Bounce Daylight into the Space While Limiting Direct Sunlight That Could Overheat the Space.

Figure 9.44 Fondazione Feltrinelli, Herzog and de Meuron, Milan, Italy Illustrating the Use of Overhangs and Coffered Windows to Provide Shade; the Cut Through Maintains an Important Urban Circulation Path While the Two Sections of Building Are Connected Below Ground.

The Fondazione Feltrinelli building in Milan, Italy houses public gallery spaces and administrative offices for the foundation. The firm of Herzog and de Meuron has conceived of the building as a long bar building with a trapezoidal plan and simple structural frame. They have designed large openings facing nominally to the south and intentionally extended the floor slabs to the exterior with intermediate exterior light shelves/overhangs to provide an integrated structure and shading system. At the latitude of 45.28°NL, the southern overhangs provide shade during the hot summer months (sun high in the sky) while allowing the sunlight to enter through the glass during winter adding warmth to the space. The long "bar" form of the building keeps the east and west façades as small as possible to keep the low morning and evening sun from penetrating deeply into the building. The shading strategy is both straightforward and yet is still compelling in terms of a modern contemporary design. By carving an urban pathway through the building, the designers have anticipated the need for public circulation and, with the exception of the moment of this photograph when the sun is shining directly into the slot, a shaded alley. Look carefully to see the operable window shades, located on the outside of the building. The are very cleanly integrated. In Figure 9.43, we can see how the design of the shade on exterior of the building as a secondary system providing user-controlled shading capacity. The shades could be thought of as extending the cutoff angle of the overhang. Figure 9.43 shows the shades on the next three levels nested within the structure of the building, yet outside the windows. This achieves a much greater shading coefficient than shading on the inside. Lastly, take a close look at Figure 9.45. Because the structure is trapezoidal at the upper window bays, the window becomes a parallelogram. The standard shades would not be able to cover the entire window if placed at the top. The architects were quite clever to mount the shades on the underside of the overhang at an angle so they "fold" down to conform to the parallelogram form of the window. Again, take a close look at Figure 9.45.

Figure 9.45 Fondazione Feltrinelli, Herzog and de Meuron, Milan, Italy Showing How the Standard Roller Shades Are Pulled Away from the Window on a Diagonal So That They Can Roll Down the Parallelogram form of the Wind; if Just Mounted at the Top, There Would Be a Light Gap.

The Asfinag building by Lorenz Ateliers in Innsbruck, Austria adopts the "fin" typology for the entire façade of the building. Here they are operable and controlled in sections so users have the opportunity to tune the angle of the fins as the sun transits the sky, balancing daylight, warmth in the winter, and avoiding glare from the low sun angles in morning and evening. Figure 9.47 is taken in the late afternoon in August and we can see the reflected glare from the sun striking the north façade.

Figure 9.46 Asfinag Office Building, Lorenz Ateliers, Innsbruck, Austria; Entry at South-Facing Façade.

Figure 9.47 Asfinag Office Building, Lorenz Ateliers, Innsbruck, Austria; Northwest Corner of Building.

Figure 9.48 Tampa Museum of Modern Art, Stanley Saitowitz, Tampa, Florida; West Façade with Large Overhang and Balcony at Civic Lawn.

Figure 9.49 Tampa Museum of Modern Art, Stanley Saitowitz, Tampa, Florida; Public Promenade, Café and Shop Along the West Façade.

Architect Stanley Saitowitz, perhaps inspired by Sarasota School architects, adopts a "super overhang" on the west façade of his museum design. He also uses an extremely deep balcony cut into the mass of the building to achieve view and shade at the interior. Since the view to the Hillsborough River is to the west, public programs such as the café and store are located there, under the shade of the super overhang. Look closely at the building skin—a perforated metal screen with an air space behind it. This is a shade screen at the scale of the building that takes the direct sun's heat, shades the building surfaces, and allows air to move through the system and carry the sun's radiant heat away.

Figure 9.50 Jepson Center, Moshe Safdie, Savanna, Georgia; Main Circulation Hall.

Figure 9.51 Jepson Center, Moshe Safdie, Savanna, Georgia; Main Circulation Hall.

Architect Moshe Safdie's Jepson Center design adapts the naturally occurring dappled shade indigenous under the oak trees of Savannah's many public squares. Breaking all the rules about glass ceilings, Safdie offers sophisticated layers of glass, structure, screens, and overhangs to filter light to shade his spaces. Safdie has managed to create the feeling of Telfair Square's oak shaded park extending into the Jepson Center.

Figure 9.52 Overhangs, Fins, Shelves, and Screens in Barcelona, Spain. RTVE Offices by Pere Joan Ravetllat (Left), CMT Headquarters by Enric Battle (Middle), and Housing by Sabate Associates.

Within a few blocks on the urban edge of Barcelona, Spain, a variety of strategies are being explored to provide shade through innovative uses of materials and clever takes on simple ideas. The CMT Headquarters takes the simple idea of the overhang or light shelf to the extreme scale of the entire building. RTVE invokes colorful hues with fin shading systems complemented by monochrome screens. Sabate uses operable wood slat screens and overhangs projected off the façade to create shade and balcony spaces.

Stefano Boeri's Vertical Garden has received many accolades internationally for his vertical landscape and using balcony/overhang combinations with vegetation as a primary shade screen. Remember that deciduous trees and plantings will lose their leaves in winter, allowing warmth from the sun to enter spaces while in the hot summer, the leafed trees will provide ample shade. Vegetation is excellent at transforming the UV light from the sun into plant material absorbing that energy rather than reflecting it onto other buildings as glass does.

Figure 9.53 Vertical Garden, Stefano Boeri, Milan, Italy as an Innovative Use of Small and Large-Scale Vegetation as a Shading Device with Deciduous Plantings That Shed Leaves in Winter Allowing Sunlight to Warm Spaces While Shading in the Hot Summer.

Solaria is an excellent example of using extended balcony spaces as overhangs shading the interior spaces. The balconies wrap façades that are oriented to the south, northeast, and northwest. The taller part of the tower in Figure 9.54 is facing northwest and the lower section of the tower is facing south. The white color of the balconies helps reflect daylight into the space while shading the direct sun that strikes the balconies. The radiant heat from the direct sun is absorbed in the outside materials while visible spectrum light is diffused into the occupied spaces.

Figure 9.54 Solaria Residential Tower, Arquitectonica, Milan, Italy, as a Creative Use of the Extended Balcony, Painted White, to Reflect and Diffuse Light Bouncing Through the Fairly Narrow Openings While Keeping the Hot Direct Sum from Penetrating and Heating the Space or Causing Direct Glare.

Mercat dels Enchants (roughly translated as Market of Charms) is a public market of repurposed things from electronics to clothes. What we might refer to as a vintage market in the United States. Vasquez Architects are quite clever here to emphasize the activity of the market through the bronzed mirrored finish of the underside of the canopy while allowing folds in the roof to make gaps and slits so skylight can enter while shading from the direct sun. The project is an amazing element in the landscape presenting the occupation of the market when viewed from blocks away.

Figure 9.55 Mercat dels Enchants, Vasquez Architects, Barcelona, Spain; Just an Overhang Without a Building.

Figure 9.56 Mercat dels Enchants, Vasquez Architects, Barcelona, Spain; Open Air Shops Under Cover.

Architect Kengo Kuma has been renowned as an emergent international practicing architect. In this showroom for stone materials in Vals, Switzerland, he expresses the slate samples as shading devices on the façade in addition to a hyper-extended overhang. The building itself resembles well stacked stones and mimicking the stacked lumber indigenous to the area. The thinness of the shade screen elements allows view, shade, and tinted light that is the color of the slate as it enters into the interior spaces.

Figure 9.57 Stone Showroom, Kengo Kuma, Vals, Switzerland.

Figure 9.58 Mercat de Santa Caterina, Barcelona, Spain, Enric Miralles; Overhang Marking the Market in the City Fabric.

Architect Enric Miralles's renovation and restoration of the historic Mercat (market) de Santa Caterina explores the concept of "canopy" and how a canopy can overhang the boundary of an historic façade (which remains) to make public space, define entry, bring in daylight, and provide shade. Miralles avoids touching the historic structure and clearly delineates the intervention, as an element that nurtures and reimagines the historic character of the market. Shade is needed, and provided by making shaded spaces for gathering, meeting, and perhaps dining adjacent to the market.

Figure 9.59 Mercat de Santa Caterina, Barcelona, Spain, Enric Miralles; Overhang Reinforcing the Central Entry of the Historic Market Façade.

Figure 9.60 Mercat de Santa Caterina, Barcelona, Spain, Enric Miralles; Overhang Defining Public Space at the Rear Entry to the Market.

Closing

Shade is a critical aspect of architecture providing comfort for occupants in warm weather and relief from the intensity of the sun. When carefully designed, shade can be controlled to provide warmth during cooler months. Shade reduces the reliance on external energy for climate control and offers an opportunity to connect interior and exterior spaces by extending the space in which people physically and visually dwell. This module has presented some practical principles, strategies, and tactics perhaps poetic ideas that can inspire the creation of shade in our architecture. As shown in the many adaptations of those basic principles in the Architectural Responses section, these principles can also enhance, and in some cases drive design concepts in positive and inspiring ways.

References

Calbó, J., Pagès, D., & González, J.-A. (2005). Empirical studies of cloud effects on UV radiation: A review. *Reviews of Geophysics*, *43*, RG2002. https://doi.org/10.1029/2004RG000155

DeKay M., & Brown G. Z. (February 3, 2014). *Sun, wind, and light: Architectural design strategies* (3rd ed.). Wiley.

Magnetic declination. (2020, September 29). Wikipedia. https://en.wikipedia.org/wiki/Magnetic_declination

Mazria, E., & Winitsky, D. (1976). *Solar Guide and Calculator*. School of Architecture and Allied Arts, Oregon University, Eugene. https://files.eric.ed.gov/fulltext/ED132716.pdf

NOAA View Global Data Explorer. (n.d.). NOAA Global Data Viewer. www.nnvl.noaa.gov. https://www.nnvl.noaa.gov/view/globaldata.html

Olgyay V. (1992). *Design with climate: A bioclimatic approach to architectural regionalism*. Van Nostrand Reinhold.

Ritchie, H., Roser, M., & Rosado, P. (2020). *Energy Production and Consumption*. Our World in Data. https://ourworldindata.org/energy-production-consumption

Image Credits

Fig. 9.1: Copyright © 1945 by Le Corbusier.

Fig. 9.6: NOAA, https://www.nnvl.noaa.gov/view/globaldata.html, 2018.

Fig. 9.20b: Copyright © by Autodesk Inc.

Fig. 9.21: Copyright © by Autodesk Inc.

Fig. 9.22b: Copyright © by Autodesk Inc.

Fig. 9.23: Copyright © by Autodesk Inc.

Fig. 9.24c: Copyright © by Autodesk Inc.

Fig. 9.24d: Copyright © by Autodesk Inc.

Fig. 9.25c: Copyright © by Autodesk Inc.

Fig. 9.26b: Copyright © by Autodesk Inc.

Fig. 9.27b: Copyright © by Autodesk Inc.

Fig. 9.28c: Copyright © by Autodesk Inc.

Fig. 9.29c: Copyright © by Autodesk Inc.

Fig. 9.43: Martin Gold, "Fondazione Feltrinelli, Herzog & de Meuron, Milan, Italy." Copyright © by Herzog & de Meuron Basel Ltd.

Fig. 9.44: Martin Gold, "Fondazione Feltrinelli, Herzog & de Meuron, Milan, Italy." Copyright © by Herzog & de Meuron Basel Ltd.

Fig. 9.45: Martin Gold, "Fondazione Feltrinelli, Herzog & de Meuron, Milan, Italy." Copyright © by Herzog & de Meuron Basel Ltd.

Fig. 9.53: Martin Gold, "Vertical Garden, Stefano Boeri, Milan, Italy." Copyright © by Boeri Studio.

Fig. 9.54: Martin Gold, "Solaria Residential Tower, Arquitectonica, Milan, Italy." Copyright © by Arquitectonica.

www.ingramcontent.com/pod-product-compliance
Lightning Source LLC
LaVergne TN
LVHW082340200926
848481LV00012B/3399